A R M Y   A N T S

THE CORNELL SERIES IN ARTHROPOD BIOLOGY

EDITED BY *George C. Eickwort*

Army Ants: The Biology of Social Predation
by William H. Gotwald, Jr.

The Tent Caterpillars
by Terrence D. Fitzgerald

ARMY ANTS

THE BIOLOGY OF SOCIAL PREDATION

William H. Gotwald, Jr.

Department of Biology
Utica College of Syracuse University

Comstock Publishing Associates A DIVISION OF

Cornell University Press | ITHACA AND LONDON

First published 1995 by Cornell University Press.

Printed in the United States of America.
Color plates printed in Hong Kong.

⊗ The paper in this book meets the minimum requirements of the American National Standard for Information Sciences—Permanence of Paper for Printed Library Materials, ANSI Z39.48-1984.

Library of Congress Cataloging-in-Publication Data

Gotwald, William H.
    Army ants : the biology of social predation / William H. Gotwald, Jr.
        p.   cm. — (Cornell series in arthropod biology)
    Includes bibliographical references and index.
    ISB 0-8014-2633-2 (alk. paper)
    1. Army ants.   2. Insect societies.   I. Title.   II. Series.
QL568.F7G66   1995
595.79′6--dc20   94-32642

For my dear friend and colleague J. K. A. van Boven, whose guidance and generosity over the years have always been seasoned with warmth, affection, and good humor.

For my mentor, William L. Brown, Jr., a myrmecologist of extraordinary accomplishment who made possible my career-long adventure with army ants.

For my daughters, Kymry and Glynis, now young women, who still regard their father's pursuit of ants with mild and bewildered amusement.

For my mother, May A. Buck, celebrating her eightieth year, who helped launch my career in biology by displaying a remarkable tolerance for my childhood preoccupation with snakes, frogs, tadpoles, and butterflies.

And for Dottie Mote Burton, whose loving commitment to the earth and its biota has challenged and inspired me to be a more strident environmentalist.

Contents

The color plates follow page 110.

Publisher's Foreword

The field of entomology is undergoing a renaissance as modern behavioral and ecological approaches are applied to the study of insects and their relatives. Recognizing the significance of this development and the need for books that consider arthropods from a modern evolutionary viewpoint, the late George C. Eickwort initiated the Cornell Series in Arthropod Biology.

The volumes in the series focus on the behavior and ecology of a particular taxon, varying in rank from class to genus. Written by scientists who are making the greatest advances in expanding our knowledge of arthropod biology, the books are comprehensive in their scope, not only detailing the subject animals' behavioral ecology but also summarizing their evolutionary history and classification, their development, important aspects of their morphology and physiology, and their interactions with the environment. Each volume can thus serve as a primer for students and scholars wishing to study that group of animals.

But beyond applying modern critical thought to the biology of arthropods, the series, we hope, will engender the great enthusiasm for entomology that George Eickwort spread to all who knew him.

Preface

What follows is a treatise on some rather extraordinary social insects. Specifically, this book is about army ants, but in its broadest sweep this chronicle explores the nature of social organization and investigates, as well, some of the determinants of social evolution. It examines the behavioral ecology of a group of insects whose social complexity and co-evolved community of symbionts and associates are without peer.

During the 1970s and 1980s, sociobiologists preoccupied themselves with the structure, function, and origin of social behavior. Most especially they sought to identify the underlying genetic bases for sociality. Social behavior, defined simply as cooperation between members of the same species that is coordinated by reciprocal communication, and its biological basis were the focus of E. O. Wilson's landmark trilogy, *The Insect Societies* (1971), *Sociobiology: The New Synthesis* (1975), and *On Human Nature* (1978). Wilson's work has had an enormous, and sometimes controversial, impact on the way we regard societies, from colonial protozoans to *Homo sapiens*.

Social evolution, Wilson (1975, p. 32) told us, "is the outcome of the genetic response of populations to ecological pressure within the constraints imposed by phylogenetic inertia." His so-called phylogenetic inertia includes the accidental predisposition of certain traits to become functional in different ways under new circumstances (such traits are termed preadaptations). In a way, phylogenetic inertia is a measure of a population's readiness to evolve. Ecological pressure, on the other hand, consists of the environmental influences, both physical and biological, that are the prime movers of natural selection and that establish the evolutionary course for a species (Wilson 1975).

The ecological pressure, or selective force, that has guided the evolution of army ants is the food on which they prey. Army ants are communal, cooperative foragers. They thereby employ a social strategy to increase their foraging efficiency. In doing so, army ants have evolved to become nature's quintessential social predators.

Communal hunting, which involves various degrees of cooperation, can be defined as the "concerted effort of at least two predators to track and hunt down prey" (Curio 1976, p. 199). The efficiency of cooperative hunting is significantly enhanced in animals with complex social behavior, especially those that hunt in packs and additionally exhibit some level of altruism. The advantage of social hunting is apparent when the prey of cooperative hunters is compared with that of solitary hunters. Among the predatory mammals, for example, those that hunt alone take animals that are usually their own size or smaller (Wilson 1975). Pack hunters such as wolves and dogs (the family Canidae), on the other hand, are able to exploit prey larger than themselves. Ipso facto, their collective efforts expand their dietary possibilities.

An example is in order. The most evolved social canid is the African wild dog, *Lycaon pictus*. Noted for its ferocity, the wild dog hunts in tight groups and relies on speed, endurance, and numbers to overtake and capture prey. Running at speeds of up to 65 km per hour, these dogs quickly catch their prey, tear it to pieces, and devour it in as few as 10 minutes. But most fascinating of all, once the satiated pack returns to the den, its members altruistically regurgitate food to the pups, their mothers, and other individuals that did not take part in the hunt. Despite the savagery of the hunt, wild dogs are "relaxed and egalitarian" in their relations with one another (Wilson 1975, p. 512). What the wild dog is to the social mammals the army ant is to the social insects.

Most striking is the cohesiveness with which the members of an army ant colony go about their business, subordinating their individuality to the demands of the collective unit. In fact, the army ant colony is not unlike a single mammalian predator. The colony thus invites comparison to a whole organism, an idea old enough to have acquired a respectable patina, a supposition recognized at least for its heuristic and "inspirational" value (Wilson 1971a, p. 317). William Morton Wheeler (1911) introduced the concept of the superorganism, comparing the worker caste to the somatic cells of a multicellular animal's body and the queen and males to its gametes. He was convinced of the truth of the concept, stating his case as follows: "The most general organismal character of the ant-colony is its individuality. Like the cell or the person, it behaves as a unitary whole, maintaining its identity in space, resisting dissolution and, as a general rule, any fusion with other colonies of the same or alien species. This resistance is very strongly manifested in the fierce defensive and offensive cooperation of the colonial personnel. Moreover, every ant colony has its own peculiar idiosyncrasies of composition and behavior" (W. M. Wheeler, 1911, p. 310).

The superorganism concept was not widely adopted, however. Wilson (1971, pp. 318–319) traced its demise, suggesting that it fell from favor because it offered no "techniques, measurements, or even definitions by which intricate phenomena in genetics, behavior, and physiology can be unraveled."

Curiously, the idea has reappeared in modified form. We are now asked to compare the ant colony to the human brain and the individual worker ant to the neuron. As Nigel Franks (1989, p. 139) pointed out, referring specifically to army ants, "the processing power of the human brain is the result of the wiring together of 100,000,000,000 neurons, one of which alone could not process even the simplest thought. Similarly, problem solving by army ant colonies is achieved through communication among some 500,000 workers, each of which has fewer than 100,000 neurons." Wilson and Hölldobler (1988, p. 66) explained that "a social organization such as an ant colony can develop properties of a problem-solving cognitive system that transcend the capacities of individual participants, even though the organization is based entirely on their communicative interactions." Regardless of the status of the superorganism concept, it is tempting to visualize foraging army ant colonies as individual predators stalking their prey within the parameters of a predator-prey system.

For twenty-five years I have enthusiastically pursued the army ant. My field studies have taken me to Africa, where I roamed the spacious savannas and verdant forests of Ivory Coast, Ghana, Nigeria, Gabon, Kenya, and Tanzania, and to Southeast Asia, where I observed army ants in the dipterocarp forests of peninsular Malaysia and in the exotic Malaysian state of Sarawak on Borneo. My field research has also taken me to Queensland, Australia, and in the New World to Mexico and Trinidad and Tobago. Moreover, I have examined type specimens in the great and not-so-great museum collections of Europe.

In the field I never tire of watching army ants; I never cease to wonder at their cohesiveness, their inexorable movement, their instinctive willfulness in taking prey, and their dramatic role in the tropical scheme of things. More than once I have attempted to crystallize my excitement in journal entries that are, by and large, too anecdotal for scientific publication.

It is my intent that this story of army ants be accurate in its science and analysis but that it be generously flavored, as well, with the awe that such creatures inspire. I do believe that awe and the curiosity it generates are driving, but underrated, forces in scientific inquiry. It is awe, too, that undergirds our respect for other living things. Without such respect, I fear that we, as a species, will continue to diminish the diversity of life on earth, and in the bargain we shall irreparably erode the quality of our own lives.

This book represents the culmination of more than twenty-five years of field and laboratory research. To name all the individuals who have facilitated and otherwise made possible my myrmecological odyssey would be a redoubtable task, far beyond the scope of these acknowledgments. A simple thank-you will have to suffice. For leaving them anonymous I apologize.

I am most indebted to those who inspired me in my continuing study of army ants, especially William L. Brown, Jr., of Cornell University, who first brought to my attention the unsolved mysteries of army ant evolution and biology. Josef van Boven, formerly of Katholieke Universiteit Leuven, Belgium, generously shared his expertise, his extensive collection of army ant queens, and his friendship. David H. Kistner, of California State University, Chico, was a frequent source of advice, encouragement, and countless specimens. My gratitude to these three gentlemen is unreserved. I also acknowledge with appreciation the words of support voiced over the years by E. O. Wilson, of Harvard University, during discussions of my research, and the special assistance I received in the field from G. R. Cunningham–van Someren, Jean Marie Leroux, Robert F. Schaefer, Jr., and the late Dennis Leston.

Special thanks are due my steadfast friend of almost thirty years, David Barr, of the Royal Ontario Museum, Toronto, who carefully read and critiqued the entire manuscript, bringing to bear in the process his consummate editorial and writing skills. Portions of the manuscript have also been critiqued by David Webb, Mark Deyrup, and David Kistner. I thank them all. Howard R. Topoff and Walter R. Tschinkel refereed the manuscript, and I value the stamp of their expertise on its final version.

Utica College, my academic home for the past twenty-six years, has provided unconditional institutional support for my research through a variety of grants and leaves. A sabbatical leave in the spring of 1990 was specifically granted for the purpose of writing this book, and financial support was provided in the summer of 1991 through a Harold T. Clark

Jr. Award. I owe much to my colleagues in the Department of Biology for their forbearance and advice during the course of my research and writing. Special thanks are due to Utica College librarians Glen E. Avery, Patricia Burchard, Eileen Kramer, Terry McMaster, and Elizabeth Pattengill, and to Interlibrary Loan Supervisor Rosemary Anguish. All have shared in (that is, did not escape) my impassioned search for esoterica. The help of Media Center personnel, most especially its director, Kathleen Randall, and William Parker, in the preparation of illustrations is gratefully acknowledged. Susan M. Dunshee, graphic designer, kindly advised me in the production of several figures. To Carol Gargas, Jean Zerbe, and Edwina Cannistra, of the duplicating department, who cheerfully tolerated my numerous requests for help, I say thank you. Deborah Paciello, secretary in the Division of Science and Mathematics, patiently typed several portions of the manuscript, and I thank her. Elizabeth M. Welch, executive secretary to the associate dean of the division, made her share of contributions to the successful completion of the manuscript as well.

I am grateful to John W. Fitzpatrick, executive director of Archbold Biological Station in Lake Placid, Florida, for permitting me to spend a part of my 1990 sabbatical at the station writing an early draft of this book. Mark Deyrup, a research biologist at Archbold, magnanimously provided space in his laboratory for me and my computer; and Fred D. Lohrer, information manager, generously helped me in my search of library resources. The many kindnesses shown me by the Archbold Biological Station staff are noted here with gratitude.

My field and museum research on Old World army ants would not have been possible without the sixteen years of continuous financial support provided by the National Science Foundation. This work was most recently underwritten by grant BSR-8403385.

Photographs and illustrations for the book were kindly provided by Roger D. Akre, Johan Billen, Ruth Chadab-Crepet, Richard J. Elzinga, Frederick R. Gehlbach, David H. Kistner, Ulrich Maschwitz, Yogendra Mahto, Keiichi Masuko, William C. McGrew, Albert Raignier, Carl W. Rettenmeyer, Kenneth V. Rosenberg, Howard Topoff, J. K. A. van Boven, Edward O. Wilson, and Allen M. Young. I am especially obliged to Carl Rettenmeyer, director of the Connecticut State Museum of Natural History, University of Connecticut, who generously supplied more than 130 photographs from which to choose illustrations for this book. The quality of his photography is renowned.

I am indebted to George C. Eickwort, original editor of the Cornell Series in Arthropod Biology and colleague and friend of twenty-eight years, whose untimely death shortly before this book went to press made the consummation of the project a somber affair indeed. His guidance and encouragement throughout the planning and writing of this book are here acknowledged. Of George's many remarkable qualities, none loomed with such singularity as his decency. I can record here only sadness knowing that he did not see this volume to completion.

To Robb Reavill, science editor at Cornell University Press, I owe an enormous thank you. Her patience in the face of my acute attacks of scriptorial procrastination is a monument to editorial stoicism. I am grateful to Helene Maddux, senior manuscript editor, for her enthusiasm and interest. And to Cornell University Press and the university itself, my graduate student home so many years ago, I say thank you.

WILLIAM H. GOTWALD, JR.

Utica, New York

1

Army Ants:
An Introduction

What's in a name? In German they are called *Wanderameisen;* in French, *fourmis voyageuses;* in Dutch, *trekmieren;* and in English, army ants. All these names refer to the way these conspicuously mobile ants move in long, orderly columns that seemingly progress with relentless, meandering intent. From a distance a column of army ants, as it clings to the contours of its diminutive landscape, looks like a graceful, undulating ribbon. Yet only a small percentage of army ant species forage or emigrate to new nest sites on the substrate surface, exposed to the threat of predators and the inquiring eyes of biologists. Regardless of the common name we apply to these ants, the vast majority lead cryptic lives within the soil and move through their milieu with the greatest secrecy. These unseen species may have their own predators, but they have yet to succumb to the insatiable curiosity of scientists. Still, what is known about army ants as notable denizens of the tropics and subtropics suggests that they play a significant role in the ecology of tropical ecosystems. Whether active at the surface or subterranean, *Wanderameisen* or *fourmis voyageuses,* army ants are important figures in the forest and savanna life of the earth's lower latitudes.

The True Army Ants

Army ants of the New and Old World tropics were once regarded as monophyletic—that is, as a taxonomic group derived from a single ancestral lineage—and were therefore placed in the single ant subfamily Dorylinae. Despite changes in our interpretation of their phylogeny,

1

these same species are still referred to collectively as the "true army ants." This is to distinguish them, as a matter of convenience and taxonomic necessity, from ant species with army ant lifeways that clearly belong morphologically to other phyletic lines. The name "army ant" is not to be confused with "legionary ant," a more restrictive term most commonly used in the literature to refer to New World army ants of the genus *Eciton,* or with "driver ant," a name applied to several rather infamous species of Old World army ants of the genus *Dorylus.*

The Army Ant Colony

The army ant colony always includes females belonging to two castes: a single fecund female referred to as the queen and sterile females called workers. Periodically, the colony also contains males (see Chapter 3).

As in all ant species, the workers are wingless. They have a pair of raised parallel ridges on the central dorsal surface of the head, called frontal carinae, that lack the lateral expansions common to other ants. The antennal insertions are thus easily visible when the army ant worker is examined from above. This feature and the fact that New and Old World army ant workers either lack eyes or have eyes reduced to a single facet makes them morphologically unique and relatively easy to separate from the workers of other subfamilies (Gotwald 1982) (Figure 1.1). But beyond these two characteristics there is much less consistency in worker morphology. The waist, or pedicel, for instance, has a single segment (called a petiole) in some species and two segments (consisting of a petiole and postpetiole) in others; and although all workers are endowed with a complete sting apparatus (Hermann 1969), some species appear incapable of stinging (Gotwald 1978b). The worker caste in all army ant genera except the Old World genus *Aenictus* is polymorphic; that is, the workers come in various sizes, forming a more or less continuous range from the smallest to the largest. As a result of allometric growth—variable rates of growth in different parts of the body—workers of dissimilar size have distinctly different proportions, especially individuals that fall at the extremes of the size range (Wilson 1971a). The largest army ant workers, notably those of the genera *Eciton* and *Dorylus,* are commonly referred to as soldiers or major workers; others are arbitrarily called media and minor workers. Workers carry out the sundry life-sustaining tasks of the colony, with individuals specializing in certain functions. When this division of labor is correlated with size, it is referred to as caste polyethism: Charles Darwin (1872, p. 205) described "working and soldier neuters, with jaws and instincts extraordinarily

Figure 1.1. Workers of the New World species *Eciton hamatum* poised above a captured wasp larva. The media worker and the soldier, with its falcate mandibles, demonstrate the spectacular nature of polymorphism in this and many other army ant species. (Photograph by Carl W. Rettenmeyer, Connecticut Museum of Natural History.)

different," in *Eciton*. When, on the other hand, the task performed is correlated with the individual's age, it is called age polyethism (Wilson 1971a).

The army ant queen is beyond doubt the most atypical queen of all ant species. More like a termite queen in her capacity to produce extraordinary numbers of eggs, the army ant queen is referred to as a dichthadiigyne because of her peculiar morphology (Wilson 1971a) (Figure 1.2). This term implies that she is blind or has reduced or vestigial eyes, that she is wingless (most ant queens initially possess wings when they emerge as adults), that she has an enlarged, single-segmented waist, and that her abdomen is greatly expanded. In all species of army ants, the queen is much larger than the workers and dramatically unlike them in appearance. In fact, the army ant queen is so unusual looking that even an experienced entomologist might not recognize her as an ant. She is the reproductive center of the colony, which cannot survive as an integrated social unit without her and will simply die, as if by maras-

Figure 1.2. *Eciton hamatum* queen and soldier. Referred to as a dichthadiigyne, the army ant queen is remarkably different from other ant queens. (Photograph by Carl W. Rettenmeyer, Connecticut Museum of Natural History.)

mus, if she is lost. Typical of all Hymenoptera, the queen's fertilized eggs produce females (most often workers), which are therefore diploid, and her unfertilized eggs yield males, which are haploid. This mode of sex determination is called haplodiploidy.

Known as sausage flies in East Africa, army ant males are robust and considerably larger than the workers. The male is winged and has mandibles that are often formidable in appearance, although the male seems unable to inflict a painful bite (Figure 1.3). Additionally, the male boasts a pair of large compound eyes; three ocelli, or simple eyes; a single-segmented waist; and genitalia that are completely retracted into the abdomen (Gotwald 1982). Males are clumsy nocturnal fliers that are commonly attracted to light and may spend much of a tropical night pointlessly thrashing about bare lightbulbs. Their wasplike habitus makes them as unlike other ants as the army ant queen. In fact, so unique is their appearance that W. M. Wheeler (1910, p. 94) conferred upon these males the name dorylaners. As with other male ants, the army ant male contributes little or nothing to the work of the colony (Wilson 1971a). These do-nothing males are practically parasitic on the colony and moreover possess a morphology adapted to their sole purpose in life, that of "flying sperm dispensers" (Wilson 1971a, p. 157).

Figure 1.3. Male of the New World genus *Neivamyrmex*. Typically army ant males have a robust, wasplike habitus. (Photograph by Carl W. Rettenmeyer, Connecticut Museum of Natural History.)

The Army Ant Adaptive Syndrome

Army ants are defined by a constellation of behavioral characteristics and an organizational structure collectively regarded as the army ant adaptive syndrome (Gotwald 1982). Foremost among these attributes are group predation and nomadism, two patterns of behavior that are the essence of what it is to be an army ant. Although neither behavior is unique to army ants, the extent to which they are combined is. Functionally related features of the syndrome include an intricate division of labor among the worker caste; huge, sometimes mammoth, colonies whose membership may reach millions; the production of new colonies through the fission of existing colonies; and a unique form of sexual selection (Franks 1989a). Each characteristic is a thread in a tightly woven net designed to capture prey not available to solitary foragers: large arthropods and social insects (Wilson 1958a).

Group predation, as defined by Wilson (1958a), includes both group raiding and group retrieval of living prey; in other words, foraging cooperatively en masse. It is this marauding behavior that prompted W. M. Wheeler (1910, p. 246), the erudite and witty founding father of myrmecology in the United States, to describe army ants as the "Huns and Tartars of the insect world." Raiding and retrieval involve different innate behavior patterns that are not always conjoined as they are in army ants. That is, although other ant species may group-retrieve their prey, these same species seldom group-raid.

Intuitively, it is easy to comprehend the expediency of foraging en masse. The advantage of large numbers of cooperating workers swarming over a cornered scorpion, with the frenzied foragers biting through the scorpion's intersegmental membranes until death is a certainty, is simple to imagine. No solitary ant forager could inflict such damage. Nor could a solitary forager successfully attack a stoutly defended ant or termite colony. Strength, as they say, is in numbers. But the relationship between group raiding and colony size is reciprocal. As much as group raiding requires a large colony size, group raiding makes a large colony size possible. Consequently, colony propagation by way of fission is the only way new army ant colonies can meet the minimal size requirement for foraging success.

Two foraging patterns, variously modified, exist in the army ants: column raiding and swarm raiding (Schneirla 1971). These patterns are species specific. A column raid has a single trunk or base column that connects the nest with the foraging arena. Smaller columns of worker ants branch from the trunk column and terminate in small groups of foraging workers. A swarm raid, a more dramatic exercise to our eyes, has a trunk column that, at its terminus, subdivides into myriad anastomosing columns, which in turn coalesce into a single advancing swarm of excited workers (Gotwald 1982). No doubt, all these foraging traits have evolved as reciprocally reinforcing features in the origin of army ant lifeways. Simply stated, army ants are successful because they have expanded their diet through cooperative foraging to include nutrient sources beyond the trophic reach of solitary hunters.

Dietary expansion through group predation is not without its potential costs. If colonies remained stationary, they might exhaust the foraging area, or trophophoric field. This is especially true for species that prey on other social insects (Gotwald 1982), whose colonies are not rapidly replaced once they are destroyed or seriously plundered.

Nomadism, an essential ingredient in the army ant adaptive syndrome, implies that army ant colonies typically emigrate from one nest site to another. Such emigrations are not in themselves unusual; some ant species emigrate two or three times in a single season (Wilson 1958a). But no other ants move with the regimented precision of the true army ants. And certainly none emigrate with the cyclic predictability seen in some of the surface-adapted New World species. In fact, T. C. Schneirla (1971) categorized army ants as belonging to one of two groups based on the regularity with which they move. In his group A he placed species that exhibit a well-defined cycle of alternating nomadic and statary phases. During the nomadic phase, a colony emigrates to a new nesting site each day after several hours of vigorous raiding. When in the statary phase, the colony raids less frequently and remains at a single nest site for the entire phase. The cyclic phenomena of these phasic species—a term I prefer to "group A" (Gotwald 1982)—are regulated through cues supplied by the brood. Schneirla's group B, which I call nonphasic, is reserved for species that conduct emigrations as single events separated by non-nomadic intervals of unpredictable length. Most army ants are nonphasic.

We can speculate with some certainty that nomadism evolved as a complement to group predation because it provides a behavioral mechanism by which colonies can periodically change trophophoric fields. The emigration patterns of species conspicuous for their surface activities are rather well documented (see, e.g., Schneirla 1971, Gotwald 1978a, Topoff and Mirenda 1980a,b, Franks 1982c, Franks and Fletcher 1983), but those of subterranean species are virtually unknown. It is tantalizing to extrapolate from the known data, but surely this is unwise.

It is apparent that a majority of army ant species are adapted to a subterranean existence. This should not surprise us if we examine the lifeways of ants in general. The evolutionary and ecological success of ants rests, in part, on the fact that they were the first group of eusocial insects to claim the soil as both home and hunting domain (Hölldobler and Wilson 1990). Subterranean species are said to be hypogaeic, and species that conduct the vast majority of their activities on the ground surface are referred to as epigaeic. These are useful terms, but in order to portray the biology of army ants precisely, the terms should be applied independently to the three components of army ant behavior: nesting, foraging, and emigration. For instance, certain African driver ants are hypogaeic nesters but epigaeic foragers.

The true army ants are not the only ant species that exhibit the army ant adaptive syndrome. Other tropical species, especially of the subfamily Ponerinae, have evolved army ant lifeways and combine, in varying degrees of complexity, group predation and nomadism (Wilson 1958a). The convergent development of group predation and nomadism in other ants suggests that the army ant adaptive syndrome is enormously successful in tropical habitats.

2

Classification and Distribution

"Classifications," wrote Ernst Mayr (1982, p. 147), perhaps the greatest evolutionist of this century, "are necessary wherever one has to deal with diversity." And Charles Darwin (1872, p. 319) wrote, "No doubt organic beings, like all other objects, can be classed in many ways, either artificially by single characters, or more naturally by a number of characters." Most simply stated, classification consists of placing individual objects into categories. Undoubtedly, the taxonomic categorization of organisms is an essential prerequisite to all empirical behavioral and ecological studies. Wilson (1971b, p. 741) pointed out that "even the most cursory ecosystem analyses have to be based on sound taxonomy." Certainly, a fundamental ingredient in understanding the significance of army ant lifeways is at least a nodding acquaintance with army ant classification and distribution and a familiarity with their journey through time.

The Family Formicidae

The ants, a group of wasplike insects that belong to the order Hymenoptera, are generally regarded as monophyletic and are therefore placed in the single family Formicidae. This family of some 12,000 to 20,000 species is usually divided into 12 or 13 subfamilies, one of which is extinct and described from fossilized specimens in amber. The subfamilies are organized into two major groups: the poneroid and formicoid complexes (Snelling 1981, Wilson 1988, Bolton 1990c). Approximately 8800 species are actually described, and these are arrayed among 297 genera (Hölldobler and Wilson 1990). A concise description of each of the subfamilies, with the exception of the Aenictinae, Dorylinae, and Ecitoninae, is provided in Table 2.1. All ants are eusocial, meaning that within the ant colony there is (1) a reproductive division of labor in which sterile individuals work on behalf of reproductive individuals, (2) cooperation in caring for the young, and (3) an overlap of at least two generations that contribute to the work of the colony (Wilson 1971a).

Ants are aculeate Hymenoptera, meaning that the females (in this case the workers) have a modified ovipositor that functions as a sting. In his cladistic analysis of the aculeate families, D. J. Brothers (1975) found the Formicidae to be close to such wasp families as the Vespidae and Scoliidae (all three families are placed in the superfamily Vespoidea) but deserving of special placement, because of their uniqueness, in an informal group that he called the Formiciformes.

The ancestry of the ants, difficult to elucidate given the relative absence of fossil evidence, was partly resolved with the discovery of the Cretaceous ant *Sphecomyrma freyi*, a species for which the subfamily Sphecomyrminae was erected (Wilson et al. 1967a,b). According to Bert Hölldobler and Wilson (1990, p. 23), "*Sphecomyrma freyi* proved to be the nearly perfect link between some of the modern ants and the nonsocial aculeate wasps." With the recent discovery of additional sphecomyrmines, Hölldobler and Wilson (1990) concluded that during the middle and late Cretaceous, a few species of this subfamily ranged widely over the northern supercontinent of Laurasia, although these ants were rather scarce compared with the ant fauna of the Tertiary (beginning 63 million years ago) and modern times.

Sometime before the Tertiary the ants underwent an adaptive radiation that was to lead to their dominance as the earth's premier social insects. In both diversity and sheer numbers the ants are unsurpassed among the social Hymenoptera. Cladistic reconstructions of the internal phylogeny of the Formicidae met with varying degrees of success until Hölldobler and Wilson (1990) reconstructed a phylogeny for the ants that included, for the first time, accumulated data on the morphology and presence/absence of ant exocrine glands. Even so, the Dorylinae (including *Aenictus*), Ecitoninae, and Leptanillinae are only tenuously represented in their phylogeny. Both Barry Bolton (1990c) and P. S. Ward (1990) produced impressive analyses of ant phylogeny that confidently include the true army ants. A phylogenetic tree synthesized from the conclusions of Hölldobler and Wilson (1990), Bolton (1990c), and Ward

Table 2.1. The subfamilies of living and fossil ants (not including the Aenictinae, Dorylinae, and Ecitoninae)

Sphecomyrminae
Extinct subfamily of Cretaceous age containing two genera, *Sphecomyrma* and *Cretomyrma* (Wilson 1987); *Sphecomyrma freyi*, discovered in amber, is the first Mesozoic ant described and is seen as a link between the nonsocial solitary wasps and modern ants.

Poneroid Complex
Cerapachyinae
A small, cosmopolitan subfamily of 197 species, in seven genera, including *Acanthostichus*, *Cerapachys*, and *Sphinctomyrmex*; this group of heavily sclerotized, elongate ants has an excursive taxonomic history, having been alternatively placed in its own subfamily or in a tribe within the Ponerinae. The cerapachyines prey on termites and ants and are related to the true army ants (W. L. Brown 1975, Bolton 1990a,c).

Leptanillinae
Containing fewer than 50 species distributed among eight genera, this small subfamily of minute ants is restricted to the Old World, with most species inhabiting the tropics and subtropics (Bolton 1990b). The majority of its species are in the genus *Leptanilla* (Baroni Urbani 1977). With the exception of *L. japonica*, little is known of their biology. They are assumed to be hypogaeic, predaceous, and group foragers; that is, they behave much like army ants. Most fascinating, the *L. japonica* queen feeds exclusively on exudations of larval hemolymph (Masuko 1987, 1989, 1990).

Myrmeciinae
A monogeneric subfamily, the *Myrmecia*, or bulldog ants, are among the most primitive of all ants. Found only in Australia and New Caledonia, they are aggressive, possess a powerful sting, nest in the soil, forage singly, and are formidable predators (Wilson 1971a); a sibling species of *M. pilosula* has a diploid chromosome number of 2, the lowest known among the insects (Crosland and Crozier 1986).

Myrmicinae
This largest and most diverse group of ants is noted for its variety in morphology and lifeways; numerous species are specialized predators, some are graminivorous, and still others, the attines, cultivate fungus gardens; many species are social parasites. The subfamily includes the huge genera *Crematogaster* and *Pheidole* (Snelling 1981).

Ponerinae
Primitive but rather diverse, universally carnivorous ants. Although many species are general predators, some species prey exclusively on isopods; others, in the genera *Proceratium* and *Discothyrea*, prey on arthropod eggs (Snelling 1981).

Pseudomyrmecinae
Represented by two genera, *Pseudomyrmex* in the New World and *Tetraponera* in the Old World. All species of these pugnacious stinging ants thus far discovered are arboreal; many species have mutualistic relationships with plants—for instance, *Pseudomyrmex* with bull's-horn acacias and *Tetraponera* with the African flacourtiaceous tree *Barteria fistulosa* (Hölldobler and Wilson 1990).

Formicoid Complex
Aneuretinae
Represented by a single living species, *Aneuretus simoni*, known only from Sri Lanka. Thought to be the direct ancestor of the Dolichoderinae, linking the dolichoderines to the nothomyrmeciines, *Aneuretus* is characterized by a mixture of specialized and primitive morphological features, including a sting (Wilson et al. 1956).

Dolichoderinae

Stingless and usually omnivorous, these ants have a rather uniform morphology. They commonly nest in the soil, although *Azteca* and *Monacis* are arboreal; *Azteca* is often associated with ant plants (myrmecophytes) such as *Cecropia* (Snelling 1981).

Formicinae

Worldwide subfamily of stingless ants, including the large genus *Camponotus* (carpenter ants). Most species are general scavenger-predators, although many rely on carbohydrates in the form of honeydew elicited from Homoptera such as aphids and treehoppers (Snelling 1981).

Nothomyrmeciinae

The single, Australian species, *Nothomyrmecia macrops*, is considered the most primitive known living ant and is sometimes referred to as a living fossil. Originally described from two specimens in 1934, it was not rediscovered until 1977. The workers are nocturnal foragers and largely nectarivorous, although they also feed on insect prey hemolymph; the chromosome number for this species (2n = 92) is the highest recorded for the Hymenoptera (Taylor 1978).

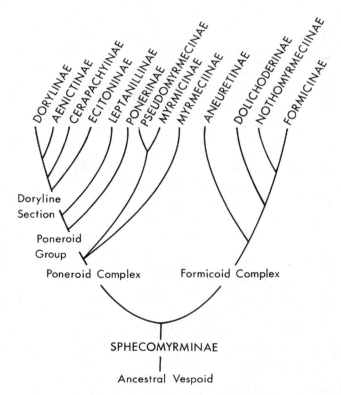

Figure 2.1. A possible phylogeny of the ants, synthesized from the analyses of Bolton (1990c), Hölldobler and Wilson (1990), and Ward (1990).

(1990) is presented in Figure 2.1. The reader is also referred to an expansive analysis of the internal phylogeny of the ants by Cesare Baroni Urbani et al. (1991). Terms applied to the external morphology of worker ants are provided in Figure 2.2. These terms are used throughout this and subsequent chapters.

A Taxonomic History

Biogeographic analyses of the pantropical distribution patterns of the true army ants (Gotwald 1979)—the original subfamily Dorylinae—culminated in the establishment of separate subfamilies for the New and Old World species: the New World forms, about 150 species, constituting the subfamily Ecitoninae, and the Old World species, perhaps 100 to 120 in number, in the Dorylinae (Snelling 1981, Watkins 1976). But in 1990 Barry Bolton, of the British Museum (Natural History), convincingly demonstrated the monophyly of the poneroid complex of ant subfamilies (Figure 2.1) and established and defined a doryline section within this complex (Bolton 1990c). A similar phylogeny of the poneroid complex was proposed by Philip Ward (1990). Based on a series of eight synapomorphies (shared derived character states), Bolton included in his doryline section the Cerapachyinae, Dorylinae, Ecitoninae, and *Aenictus*, which he elevated to subfamily status (Aenictinae). Earlier he (1990a) had reinstated the subfamily Cerapachyinae from its position within the Ponerinae and (1990b; see also Hashimoto 1991) persuasively argued that the Leptanillinae, which he reorganized and redefined, should be uncoupled from its traditional association with the army ants. In a study of sting apparatus morphology, Charles Kugler (1992) concurred that there was no evidence supporting the Leptanillinae as a sister group of the dorylines and ecitonines. Bolton's conclusions are adopted here, although the "true" army ants (Aenictinae, Dorylinae, and Ecitoninae) remain the focus of attention.

Carolus Linnaeus (Linné 1764), Swedish botanist and originator of modern classification, described the first army ant but assumed it was a wasp and placed it in the genus *Vespa*. In doing so, he unwittingly started what was to become a convoluted taxonomic history, compounded by the errors of subsequent taxonomists, forming a Gordian knot that was not to be cut for eight decades. Linnaeus described *Vespa helvola* on the basis of two male specimens from the Cape of Good Hope belonging to a species now placed in the Old World army ant genus *Dorylus* (Plate 1A). Central to his taxonomic misplacement of this species

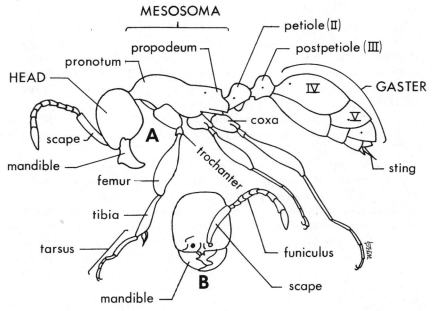

Figure 2.2. Worker of the African army ant *Aenictus decolor*, with the principal external morphological features identified (A, habitus; B, head, right antenna omitted).

were two factors that should garner at least a modicum of sympathy for Linnaeus: army ant males are indeed wasplike and unlike the vast majority of other ant males, and the males he examined were not collected with members of the worker caste. The fact that army ant males, queens, and workers are only rarely collected together explains much of the taxonomic confusion that has obfuscated our understanding of their diversity. And so Linnaeus (1764, p. 412) described his new species: "Vespa testacea, thorace hirsuta, pedibus ferrugineis, femoribus compressis" (wasp reddish brown, thorax hairy, feet rust-colored, femora compressed).

Apparently uncomfortable with his original placement of *helvola* in the genus *Vespa*, Linnaeus (Linné 1767), in the famous twelfth edition of his *Systema naturae*, transferred the species to the wasp genus *Mutilla*, a group commonly referred to now as the velvet ants. Although the brightly colored females of this genus are wingless and antlike, they are most definitely wasps, not ants. The taxonomic odyssey continued when J. C. Fabricius (1793), a Danish student of Linnaeus and the foremost entomologist of the eighteenth century, created for *helvola* the genus *Do-*

rylus. The taxonomic history of *Dorylus* before 1840 is a meandering stream cluttered with the flotsam of esoteric debate. In that year, J. O. Westwood (1840a,b), an active archaeologist as well as a prodigious entomologist and Hope Professor in Oxford, described the first Old World army ant worker and insisted that his species, *Typhlopone fulvus,* was indeed an ant. The specimens of this Old World species on which Westwood based his description, however, were found in sugar casks mistakenly listed as coming from South America or the West Indies, and biogeographic confusion was thus introduced to the simmering taxonomic stew.

Coincidentally, in 1840 W. E. Shuckard published the first monographic review of the army ants—known only from males and, of course, not yet recognized as ants. Although Shuckard (1840, p. 193) noted similarities between the ant genus *Ponera* and the dorylines (the family Dorylidae had been created for *Dorylus* and the New World genus *Labidus* by William Leach in 1815), he insisted that the dorylines were distinct from the ants and might even be "parasites upon the Social Ants." He included in his taxonomic treatment several wingless forms that he deductively assumed were the females of *Labidus* and *Dorylus,* and he subsumed in this group Westwood's *Typhlopone.* Shuckard was indeed correct in his assumption that the wingless forms were dorylines, but he nevertheless precipitated an abstruse debate. Westwood maintained that his *Typhlopone* specimens were ants, while Shuckard opined that this was not possible because they belonged to the non-ant family Dorylidae. Of course, in truth they *were* ants—members of the army ant worker caste. Interestingly, at least four species of New World army ants had been described from the worker caste before 1840 and were recognized as ants from the start.

The army ant mystery was eventually solved. In 1849, the Reverend Thomas S. Savage, a medical missionary in West Africa, observed wingless *Dorylus* males running in a column of workers of the army ant *Anomma rubella* (in reality a species of *Dorylus* commonly referred to, with other similar species, as driver ants). At first, Savage (1849, p. 201) supposed the males to be captives, but he thought better of the idea after a small field experiment: "I was soon convinced that they belonged to the drivers, and proceeded to test the truth of the conclusion, I took one or two from the lines to a distance of six and ten feet. They seemed at once to miss their companions, and manifested great trepidation, and made continuous efforts to find a way of return. At last they reached the lines and instantly resumed their places, displaying at the same time decided gratification."

Savage submitted his written observations for publication to the Academy of Natural Sciences in Philadelphia, along with a sample of specimens. The committee that reviewed his manuscript and examined the specimens noted that there was "sufficient evidence that *Anomma* . . . is another condition of *Dorylus*. . . . which must take its place among the Formicidae" (Savage 1849, p. 201). In other words, *Dorylus* was discovered to be an ant.

Although the pieces of the taxonomic jigsaw puzzle were falling into place, no fertile female, or queen, had yet been found. In 1863 the puzzle was essentially completed when Adolph Gerstaecker described the first queen, *Dichthadia glaberrina*. Although his description was based on a single isolated specimen, he conjectured that it probably was the female of *Dorylus*. The taxonomy of the Ecitoninae and Dorylinae still suffers from the lack of associated males, workers, and queens in many of the species. Certainly, numerous species are based on the descriptions of isolated—that is, unassociated—males, workers, and queens, creating numerous synonymies that await discovery.

Important monographic works have broadened our understanding of army ant diversity and have provided a basic level of taxonomic stability. Myrmecologist Carlo Emery, for example, revised the genus *Dorylus* in 1895 and summarized the classification of the entire subfamily Dorylinae in 1910. American entomologist Marion R. Smith (1942) considered the taxonomy of *Neivamyrmex* army ants in the United States, and Fr. Thomas Borgmeier (1953, 1955), of Brazil, produced a monumental and lasting revision of the New World army ants. Julian Watkins (1976) supplemented Borgmeier's work, and E. O. Wilson (1964) revised the Indo-Australian species of *Aenictus* and *Dorylus*, genera still in need of a complete taxonomic revision. Phenetic studies of these two genera in preparation for such a revision have yielded valuable insights into the current subgeneric classification of *Dorylus* and phyletic interpretations of *Aenictus* (Gotwald and Barr 1980, 1987; Barr and Gotwald 1982; Barr et al. 1985). Undoubtedly, Bolton's (1990c) phylogenetic analysis of the army ants has brought new clarity to our understanding of their phylogeny. The army ants are currently ordered among the following subfamilies, genera, and subgenera:

> Subfamily Aenictinae (Old World)
>   Tribe Aenictini
>   Genus *Aenictus*
> Subfamily Dorylinae (Old World)
>   Tribe Dorylini
>   Genus *Dorylus*

16

Subgenus *Alaopone*
Anomma
Dichthadia
Dorylus
Rhogmus
Typhlopone
Subfamily Ecitoninae (New World)
Tribe Cheliomyrmecini
Genus *Cheliomyrmex*
Tribe Ecitonini
Genus *Eciton*
Labidus
Neivamyrmex
Nomamyrmex

The Genera and Subgenera

Although the vast majority of army ant species are tropical and are therefore found between the Tropics of Cancer and Capricorn, a few species are clearly adapted to temperate conditions, and some can survive rather harsh winters. Because most species are subterranean, our knowledge of army ant biology has a rather random quality. Too often our observations have depended on the serendipitous discovery of colonies that are normally concealed within the dark depths of the soil. Only the surface-adapted species have been systematically investigated. Below are short synopses of each genus that include essential facts about morphological characteristics, distribution, and biology. Component features of army ant lifeways are examined in greater detail in subsequent chapters.

Subfamily Aenictinae

Aenictus
Characterized by having the most diminutive and least conspicuous workers, *Aenictus* is strongly represented in the Indo-Australian region with more than 40 species (Wilson 1964, W. H. Gotwald, Jr., unpubl. data) but only moderately so in Africa with approximately 15 species. In all species the workers are monomorphic, eyeless, have an undivided mesosoma, possess a two-segmented waist, and have a well-developed sting. Although the genus is not divided into subgenera, four phenetic subgroups are recognizable (Gotwald and Barr 1987). Two of these groups, the African species and the Asian forms once referred to as *Typhlatta*, are especially distinct. The vast majority of species are de-

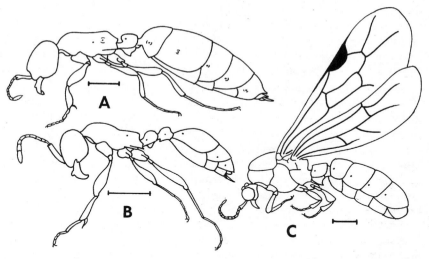

Figure 2.3. Queen (A) and worker (B) of *Aenictus decolor*; male of *A. hamifer* (C). Each scale equals 1 mm. (Queen and worker redrawn, by courtesy of The Entomological Society of Washington, from Gotwald and Leroux 1980; male redrawn from Emery 1910.)

scribed from either isolated workers or males, resulting in two classifications of *Aenictus* (Wilson 1964). Both workers and queens are known in only 8 species, 4 each from Asia and sub-Saharan Africa (Wilson 1964, Gotwald and Cunningham–van Someren 1976, Gotwald and Leroux 1980, Campione et al. 1983). The queens are typical dichthadiigynes. Suturing in the mesosoma of African queens is reduced (derived); that of Asian queens is rather complex (primitive). In fact, the morphological differences between the African and Asian queens prompted W. M. Wheeler (1930) to suggest that the Asian species belonged to a distinct genus. Males of *Aenictus* have large compound eyes and well-developed ocelli. They are most easily distinguished from *Dorylus* by their much smaller size (body length, exclusive of mandibles, is usually less than 8.5 mm) (Gotwald 1982) (Figure 2.3).

Most species of *Aenictus* are specialized predators of other ants, especially of the immature stages (Gotwald 1976, Rosciszewski and Maschwitz 1994). Only the Asian species *A. gracilis* and *A. laeviceps* are general predators, taking a wide variety of invertebrate prey. These two species are suspected of collecting sweet secretions from plant nectaries, a significant departure from the observed foraging behavior of other army ants (Schneirla and Reyes 1966). In another departure from the army ant diet, the African species *A. eugenii* reportly tends and collects honeydew from the homopteran *Pseudococcus* (Santschi 1933). Most species are generally hypogaeic in their activities, including nesting, with the exception again of *A. gracilis* and *A. laeviceps*, which forage and emigrate on the substrate surface. When nomadic, they also form temporary disk-shaped bivouacs instead of subterranean nests. The bivouac is simply a thin cluster of workers on the ground, surrounding the brood and queen (Schneirla and Reyes 1969). A single *A. laeviceps* colony may contain as many as 60,000 to 110,000 workers (Schneirla and Reyes 1966). The frequency and cyclical nature of emigrations in *Aenictus* are basically unknown. Only the epigaeic species *A. gracilis* and *A. laeviceps* are known to exhibit a functional cycle of alternating statary and nomadic phases (see Chapter 3) with predictable emigrations. These two species are, however, anomalous compared with other *Aenictus* species.

Aenictus is recorded from India, various countries of Southeast Asia, and southern China. It also has been collected in Australia (Queensland and New South Wales), the Philippines, Sri Lanka, New Guinea, Borneo, Java, Aru, Japan, and Taiwan (Wilson 1964, Terayama 1984). The genus is found throughout much of Africa and ranges through the Middle East to the Far East, but no species appear to be shared in common between the Asian and African populations. As mentioned above, the Asian and African queens have distinct morphologies, which suggests a relatively long period of isolation from one another (W. M. Wheeler 1930). Ecologically, *Aenictus* is the equivalent of the New World genus *Neivamyrmex;* that is, it is the most widely distributed Old World genus and has dispersed into temperate habitats as well. It has been collected, for instance, in Afghanistan, along the mountainous Pakistan frontier, near Peshawar (Pisarski 1967), and in Armenia (Arnoldi 1968) (Figure 2.4).

Subfamily Dorylinae

Dorylus
The workers of all species of *Dorylus* are polymorphic, lack vestiges of eyes, and possess a uninodal waist and a well-developed sting. The mandibles of *Dorylus* workers exhibit a diverse morphology. Even within a single species, the mandible undergoes a gradual but continuous transition in shape from the smallest to the largest workers, producing mandible morphologies variously adapted to foraging and other tasks (this is true of all polymorphic species) (Gotwald 1978b). *Anomma* soldiers have falcate, sharply pointed mandibles, which clearly are adapted to a defensive function (Wilson 1971a). The mesosoma is divided in half dorsally and laterally by sutures, and this character alone

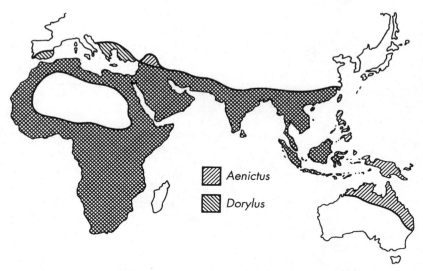

Figure 2.4. Known distribution of the genera *Aenictus* and *Dorylus* based on literature cited in the text.

distinguishes *Dorylus* workers from the ecitonines and *Aenictus* (Figures 2.5–2.7). The most diagnostic feature of the worker, however, is a circular concavity impressed upon the pygidium, or last dorsal sclerite of the gaster. This pygidial impression may be circumscribed by either a sharp, elevated margin (as in the subgenera *Dorylus* and *Anomma*) or simply a rounded border, but in all cases the impression is armed on each side with a spine of unknown function (Gotwald 1978b). *Dorylus* queens are the largest ants known. They are blind and their mandibles are linear, curved apically, and without teeth along the inner border. The single-segmented petiole may be enlarged and armed with caudally directed horns (Figure 2.5). Males of *Dorylus* are among the largest and most robust of all army ant males (Gotwald 1982) (Figures 2.5–2.7). Phenetic studies of *Dorylus* workers, males, and queens have identified three discrete species groups: (1) *Dorylus (sensu stricto)*, plus *Anomma, Dichthadia,* and *Typhlopone;* (2) *Rhogmus;* and (3) *Alaopone* (Gotwald and Barr 1980, Barr and Gotwald 1982, Barr et al. 1985).

The vast majority of *Dorylus* species are hypogaeic in both foraging and nesting. *Anomma* species are the exception, conducting both raids and emigrations on the substrate surface. Subcaste polyethism—division of labor among workers that is correlated with size—is documented for *Anomma* (Gotwald 1974a). Although subterranean species tend to be

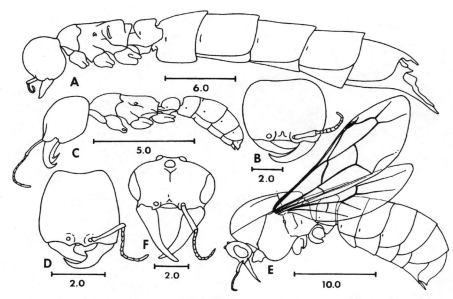

Figure 2.5. The queen (A, legs omitted), head of the queen (B), major worker (soldier) (C, legs omitted), head of the major worker (D), male (E, legs omitted), and head of the male (F) of *Dorylus* (*Anomma*) *molestus*. The subgenus *Anomma* is distinguished by its highly polymorphic workers, immense queens, and robust, wasplike males. Scales in millimeters. (Reprinted, by permission of Academic Press, from Gotwald 1982.)

trophic specialists that feed exclusively on other ants or termites, a few species of *Anomma* are notorious generalists. These same species may have colonies with as many as 10 million to 20 million workers (Raignier and van Boven 1955). Little is known of their emigration behavior and the cyclicity and synchrony of brood production. No species yet examined has a reproductive cycle of alternating statary and nomadic phases. Indeed, *Anomma* species are known to emigrate at irregular intervals and to remain in their subterranean nests for extended and unpredictable periods of time (Gotwald and Cunningham–van Someren 1990).

Essentially, *Dorylus* is an African genus. Only 4 of its 30 to 40 species are found in India and tropical Asia, although the subgenus *Dichthadia* is endemic to the Oriental region. The Asian species include the sole species of *Dichthadia*, 2 of the subgenus *Alaopone*, and 1 of *Typhlopone*. The remaining subgenera (*Anomma*, *Dorylus*, and *Rhogmus*) are more or less endemic to the African continent and are not found in Asia (Wilson 1964, Gotwald 1982). The subgenus *Anomma*, whose taxonomic status is

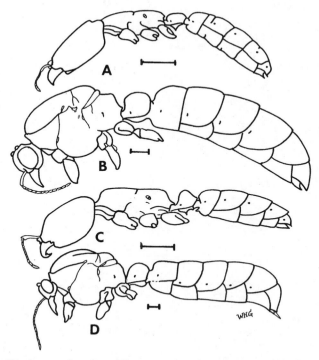

Figure 2.6. Workers and males of the genus *Dorylus* (legs omitted, wings of males omitted). Major worker (A) and male (B) of the subgenus *Alaopone*, and major worker (C) and male (D) of the subgenus *Dichthadia*. Each scale equals 1 mm. (Workers redrawn, by permission of the Entomological Society of America, from Gotwald and Barr 1980; males redrawn, by permission of the National Research Council of Canada, from Barr and Gotwald 1982, *Canadian Journal of Zoology* 60: 2652–2658.)

in doubt (Barr et al. 1985), includes the species known as driver ants and has the most limited distribution in Africa. *Anomma,* the ecological and behavioral analogue of *Eciton,* appears to be less successful in xeric and cool habitats than the other African subgenera (Gotwald 1982). Like *Aenictus, Dorylus* is apparently continuously distributed across the Middle East, although Africa and Asia do not share any species. *Dorylus* has not dispersed to temperate habitats as has *Neivamyrmex* in the New World, although one species, *Dorylus* (*Typhlopone*) *fulvus,* is recorded from Spain and Yugoslavia (Collingwood 1978, Agosti and Collingwood 1987) (Figure 2.4).

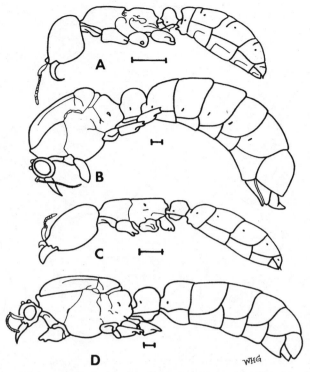

Figure 2.7. Workers and males of the genus *Dorylus* (legs omitted, wings of males omitted). Major worker (A) and male (B) of the subgenus *Rhogmus*, and major worker (C) and male (D) of the subgenus *Typhlopone*. Each scale equals 1 mm. (Workers redrawn, by permission of the Entomological Society of America, from Gotwald and Barr 1980; males redrawn, by permission of the National Research Council of Canada, from Barr and Gotwald 1982, *Canadian Journal of Zoology* 60: 2652–2658.)

Subfamily Ecitoninae

Cheliomyrmex

Of the five ecitonine genera, *Cheliomyrmex* is the most decidedly tropical and restricted in distribution. It comprises a perplexing and biologically cryptic group of five species that together constitute the tribe Cheliomyrmecini. The workers' morphology is also the most distinctive of the New World genera—so much so that W. M. Wheeler (1921) characterized the tribe as an ancient group and the most generalized of all the tribes of the Dorylinae (*sensu lato*).

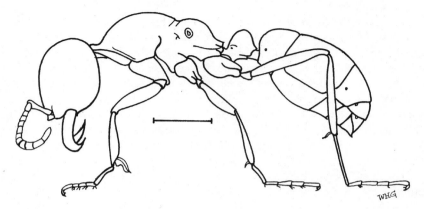

Figure 2.8. Major worker of *Cheliomyrmex morosus*. Scale equals 1 mm. (Redrawn, by permission of the Entomological Society of America, from Gotwald and Kupiec 1975.)

Cheliomyrmex workers are moderately polymorphic, falling somewhere between the monomorphism of *Aenictus* and the strongly polymorphic *Eciton,* and, unlike other Ecitoninae, are eyeless (Gotwald and Kupiec 1975). The most striking difference between this genus and the other ecitonines is in the condition of the waist, which is uninodal in *Cheliomyrmex* and binodal in the rest of the subfamily (Figure 2.8). Like other ecitonines, *Cheliomyrmex* has a suturally undivided thorax, a condition that clearly separates the ecitonines from *Dorylus,* with its bipartite thorax (Reid 1941). The queen of *Cheliomyrmex* is unknown; males, on the other hand, are not uncommon in collections. The rather hirsute male is known for four species, two of which are based solely on the male (Watkins 1976). W. L. Brown, Jr., and W. L. Nutting (1950) concluded that, among the army ants, the *Cheliomyrmex* wing is the most primitively veined.

Cheliomyrmex, clearly hypogaeic in habit, has been collected so infrequently that its biology is virtually unknown, although C. *morosus* is fairly commonly encountered, especially in the vicinity of Orizaba, Cordoba, and Atoyac in the state of Veracruz, Mexico (Borgmeier 1955). I observed one colony, discovered beneath a log in a pasture near Atoyac, the type locality for the species, but was unsuccessful in my search for the queen and brood. Although some tunnels descended to a depth of about 15 cm, most were located just beneath the surface. When disturbed, the workers quickly sought refuge in the tunnels and could no longer be found. In defending itself, the worker bites and stings simultaneously. The sting is painful, rather like a mild bee sting, and produces a small, circular welt about 7 mm in diameter (Gotwald 1971). W. M. Wheeler described the sting of *C. megalonyx* as painful and found this species to be strongly photophobic. One colony that he found beneath a pile of logs emigrated after being disturbed. Its columns moved beneath leaves, sticks, and boards, where such cover was available. Open spaces were crossed through galleries which the workers constructed of small particles of earth (Wheeler 1921). Such galleries are commonly built by various species of the genus *Dorylus*. Only larvae were collected from this emigration column, suggesting that the brood are produced synchronously.

Because *Cheliomyrmex* is only rarely collected, its distribution may be much wider than it is currently believed to be (Figure 2.9).

Eciton

The 12 *Eciton* species (with numerous named subspecies), all noted for their dramatic and conspiuous foraging and emigration behavior, are limited to tropical habitats. Two, *E. hamatum* and *E. burchelli*, are not only the most studied and best known of the army ants but may rank among the most intensely investigated of all ant species. T. C. Schneirla devoted much of his distinguished career to them (see Schneirla 1971), and they have garnered the attention of a host of other researchers as well, including Carl W. Rettenmeyer and Nigel R. Franks, who have added significantly to our knowledge of these species.

Eciton species are highly polymorphic, as evidenced in the size differential between the smallest and largest workers. For *E. burchelli*, this differential is 8.1 mm (as compared to 0.5 mm for the monomorphic *Aenictus gracilis*) (Gotwald and Kupiec 1975). The workers have compound eyes reduced to a single facet. *Eciton* soldiers have spectacular falcate (hook-shaped) mandibles that are diagnostic of the genus (except *E. rapax*) (Figure 2.10). The mesosoma is suturally undivided, the waist has two segments, and the sting is strongly developed. The queens also possess reduced compound eyes. The queen's propodeum (posterior end of the mesosoma) and petiole are armed with caudally directed horns (Gotwald 1982) (Figure 2.10). In two observed *Eciton* matings, the males grasped a petiolar horn with their mandibles during copulation (Schneirla 1949) (see Figure 4.16).

Both *E. hamatum* and *E. burchelli* are truly epigaeic, and both are probably the most highly specialized of the 12 or so *Eciton* species (Rettenmeyer 1963b); that is, they are to *Eciton* what *A. gracilis* and *A. laeviceps* are to *Aenictus*—atypical. They are generalist predators and exhibit a functional cycle of alternating statary and nomadic phases related to the synchronized development of the brood (see Chapter 4). *E. hamatum* colonies may con-

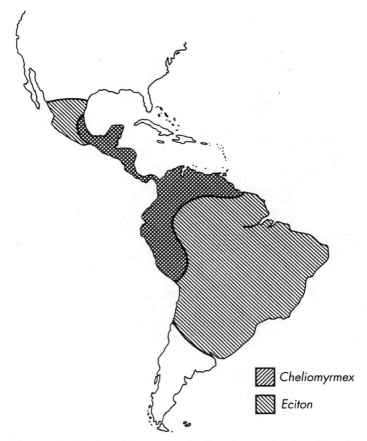

Figure 2.9. Known distribution of the genera *Cheliomyrmex* and *Eciton*. (Based on Watkins 1976, 1985.)

tain 150,000 to 500,000 workers, those of *E. burchelli* 500,000 to 2 million workers (Rettenmeyer 1963b). In addition to foraging and emigrating on the surface, they form temporary above-ground nests called bivouacs, large suspended clusters of workers in which the brood and queen are sequestered (Rettenmeyer 1963b, Schneirla 1971). Other species of *Eciton* prey primarily on ants, and some, like *dulcius* and *vagans*, are rather subterranean and may not engage in the strict alternation of statary and nomadic phases seen in *burchelli* and *hamatum* (Rettenmeyer 1963b).

Eciton ranges from Mexico south through Brazil and into northern Argentina (Figure 2.9).

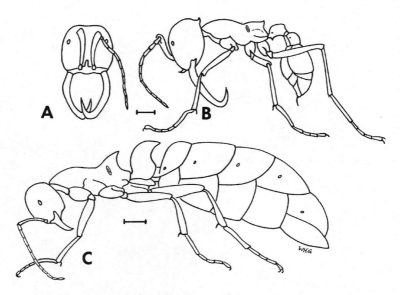

Figure 2.10. Major worker (soldier) of *Eciton hamatum* (A, head; B, habitus) and queen of *E. dulcius* (C). The genus *Eciton* includes species noted for their surface-active lifeways. Scale equals 1 mm. (Worker redrawn, by permission of J. F. Watkins and the Kansas Entomological Society, from J. F. Watkins 1982.)

Labidus

The eight species, four of which are known only from males, are primarily subterranean. The workers are polymorphic, have reduced compound eyes, have a mesosoma like that of *Eciton* but without the posterior, toothlike projections common to *Eciton*, and a two-segmented waist (Figure 2.11) (Gotwald 1982). There is no clear morphological distinction between the largest workers, referred to as soldiers, and the smaller workers, as there is in *Eciton*. One species, *L. praedator*, is reported to have strong bites but a sting that usually does not penetrate human skin (Rettenmeyer 1963b, p. 404). The queen is known for two species and can be distinguished from the *Eciton* queen by her propodeum, which is not adorned with horns. The male closely resembles other ecitonine males, including *Cheliomyrmex*, and can be separated from them only through rather subtle characters.

The biology of two species, *L. praedator* and *L. coecus*, is at least partially understood. An *L. praedator* colony may contain in excess of a million workers, perhaps the largest among the ecitonines (Fowler 1979).

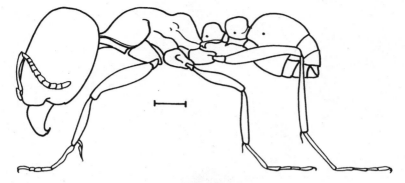

Figure 2.11. Major worker of *Labidus coecus*. Scale equals 1 mm. (Redrawn from M. R. Smith 1965.)

This species is a general predator but also takes unusual foods (for army ants) such as sugar, crushed pineapple, and boiled rice (Borgmeier 1955). It forages both day and night, conducting its raids largely underground and beneath leaf litter. *L. praedator* nests in the soil, especially in preformed cavities such as abandoned leaf-cutter ant nests, and may remain at one nest site for months at a time (Fowler 1979).

The genus *Labidus* is distributed from Argentina to Oklahoma, with a single species, *L. coecus*, accounting for the extremes reached both north and south of the equator (Watkins 1976) (Figure 2.12).

Neivamyrmex

Using species diversity as a measure of evolutionary success, one would judge *Neivamyrmex* as the most successful of all army ant genera. Compared with other ecitonines, its morphology is undistinguished. The workers possess reduced eyes and rather unremarkable mandibles; they have a two-segmented waist and a well-developed sting (Figure 2.13). The workers differ, however, from all the other Ecitoninae by having tarsal claws without teeth. The *Neivamyrmex* queen also possesses simple tarsal claws, which separates her from queens of the other genera (Figure 2.13). Males, on the other hand, differ from other genera less dramatically (Gotwald 1982) (Figure 2.13).

The most intensely studied species in this genus is *N. nigrescens*, and much of what we know about this species comes from the exemplary work of Howard Topoff, a former student of T. C. Schneirla, and his students and co-workers. *N. nigrescens* colonies contain 80,000 to 140,000 workers (Schneirla 1958), which nest in the soil (Rettenmeyer 1963b). The species exhibits a functional cycle of alternating nomadic and statary

Figure 2.12. Known distribution of the genera *Labidus* and *Nomamyrmex*. (Based on Watkins 1976, 1985.)

phases of 17 to 20 days each. The nomadic phase is characterized by a high level of colony activity with nightly raids that end in emigration to a new nest site. During the statary phase, there is less rigorous raiding and a commensurate absence of emigrations. The cycle is regulated by the relationship between the colony's developing brood and the adult workers (Topoff 1975, p. 38); the larvae stimulate raids as well (Topoff and Mirenda 1980a,b). In the desert grasslands of Arizona and New Mexico, *N. nigrescens* preys exclusively on termites and immature stages of ants (Mirenda et al. 1980). Aspects of the biology of other species were documented by Carl Rettenmeyer (1963b), among others.

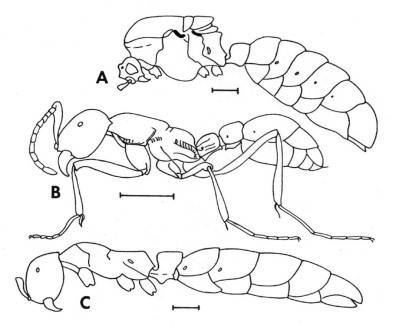

Figure 2.13. Male (A, legs and wings omitted), worker (B), and queen (C, legs omitted) of *Neivamyrmex nigrescens*. Each scale equals 1 mm. (Worker redrawn from M. R. Smith 1965; queen and male redrawn, by permission of J. F. Watkins and the Kansas Entomological Society, from Watkins 1972.)

The genus *Neivamyrmex*, to which belong the vast majority of ecitonine species (more than 120), is the most wide ranging of all army ants, extending to 40° latitude on either side of the equator. *Neivamyrmex* species are reported from locations as far north as Iowa, Illinois, and West Virginia (Watkins 1976, 1985; DuBois 1988) (Figure 2.14).

Nomamyrmex

Two species, *N. esenbecki* and *N. hartigi*, constitute this genus. The taxonomy of *esenbecki*, however, is confusing. No fewer than 14 names have been applied to its various forms. Four subspecies are now recognized (Watkins 1977), but their biological validity as distinct geographical races could be challenged. The workers are not unlike those of *Eciton*, although they lack the flamboyant mandibles of the *Eciton* soldiers (Figure 2.15). An *esenbecki* queen was collected by Rettenmeyer (1963b) in 1956 and described by Borgmeier (1958), who noted that the habitus of the *Nomamyrmex* queen is generally rather similar to that of the *Eciton* queen. The males are typical ecitonines (see Watkins 1976).

Figure 2.14. Known distribution of the genus *Neivamyrmex*. (Based on Watkins 1976, 1985.)

What is known of the biology of *Nomamyrmex* comes primarily from observations of *esenbecki,* so seldom has *hartigi* been seen. Rettenmeyer (1963b) found that 80 to 90 percent of *esenbecki*'s prey consisted of ant larvae and pupae, although Borgmeier (1955) observed that ants of this species captured many adult ants as well. Henry Walter Bates (1863, p. 375) noted that "this ant goes on foraging expeditions like the rest of its tribe, and attacks even the nests of other stinging species (Myrmica), but it avoids the light, moving always in concealment under leaves and fallen branches. When its columns have to cross a cleared space, the ants construct a temporary covered way with granules of earth, arched over,

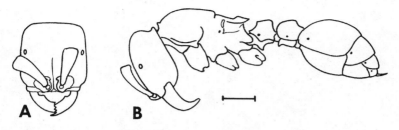

Figure 2.15. Worker of *Nomamyrmex esenbecki wilsoni* (A, head, antennae incomplete; B, habitus, legs omitted). Scale equals 1 mm. (Redrawn, by permission of J. F. Watkins and the Kansas Entomological Society, from Watkins 1982.)

and holding together mechanically; under this the procession passes in secret, the indefatigable creatures repairing their arcade as fast as breaches are made in it." Rettenmeyer (1963b) also noticed that this species builds canopies over its foraging columns during daylight raids. Although many colonies of this species have been observed, the nest, which is subterranean, has never been found. Evidence suggests that *esenbecki* may have an activity cycle like that of *Eciton hamatum* (Rettenmeyer 1963b).

N. *hartigi* has a slightly smaller range than *esenbecki,* from Mexico City to southern Brazil. *N. esenbecki,* in its various forms, ranges south from Austin, Texas, to southern Brazil (Watkins 1977) (Figure 2.12).

Origin of Army Ants

The Fossil Record

The origin of army ants is problematic, for little direct evidence of their ancestry exists. An extinct species of *Neivamyrmex* recently found in Dominican amber (i.e., from Hispaniola), is the first army ant in the fossil record (Wilson 1985b). Although the exact age of the amber is not known, most deposits probably originated no later than the early Miocene (Figure 2.16). Wilson (1985a), who analyzed the Dominican amber ant fauna, favored a minimal age of about 20 million years. In truth, the discovery of this extinct army ant reveals more about West Indian biogeography than it does about army ant origins. Because army ants are not presently found in the Greater Antilles (Cuba, Jamaica, Hispaniola, and Puerto Rico), and because they do not disperse well across water, the presence of *Neivamyrmex* in the Dominican amber suggests that these islands were larger and extended closer to Mexico during the middle and late Tertiary than is presently the case (Wilson 1985b). This conclu-

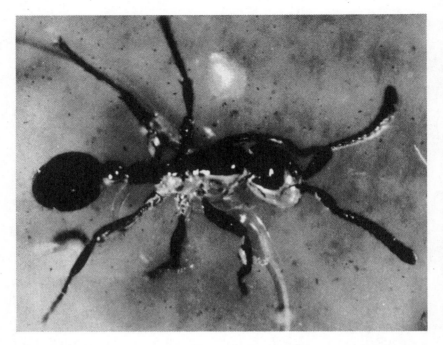

Figure 2.16. The extinct species *Neivamyrmex ectopus* in Dominican amber, the first army ant found in the fossil record. (Photograph by F. M. Carpenter, courtesy of E. O. Wilson.)

sion is apparently consistent with views based on geological and paleobotanical studies.

The Role of Plate Tectonics

Army ant phylogeny must be inferred from comparative zoogeographical, morphological, and behavioral studies. Given certain assumptions, present-day distribution patterns of army ants can be correlated with the movement of continents, and phyletic conclusions can be drawn. The theory of continental drift posits that today's continents once formed a single land mass, Pangaea. During the late Triassic to mid-Jurassic (about 180 million years ago), Pangaea began to split into a northern and a southern cluster of continents, called Laurasia and Gondwana, respectively (Dietz and Holden 1970). Eventually, Laurasia and Gondwana themselves fragmented into the northern and southern continents. By the end of the Cretaceous (63 million years before the present), South America and Africa were widely separated by 3000 km of South Atlantic Ocean, and early in the Cenozoic (which immediately followed the Cretaceous), the North Atlantic rift completed the breakup of Laurasia, separating North America from Eurasia (Dietz and Holden 1970). Even Arabo-Africa and Eurasia were divided from one another during much of the Mesozoic and Tertiary by the pre-Mediterranean Tethys Sea (Cooke 1972). Based on the estimated antiquity of the ants in general (see Wilson et al. 1967a,b) and the diversity of the Oligocene (36 million to 25 million years before the present) ant fauna, Gotwald (1977, 1979) hypothesized a late Cretaceous, possibly early Tertiary, origin for the Dorylinae (including *Aenictus*) and Ecitoninae. Schneirla (1971) postulated a similar origin for these ants. By the end of the Cretaceous, the three tropical areas in which the true army ants are currently found were already separated by substantial bodies of water.

Because *Dorylus, Aenictus,* and the Ecitoninae probably arose after the breakup of Gondwana and Laurasia, which prevented effective dispersal, these three groups do not share a common ancestry (Gotwald 1977, 1979). That is, they arose convergently on three occasions at three separate tropical loci: the Ecitoninae in South America, *Dorylus* in Africa, and *Aenictus* in Laurasia. The probability is low that the army ants, poor dispersers, originated in a single place and then crossed significant ocean barriers. What makes army ants such poor dispersers? Quite clearly, they are restricted by the fact that the queen is wingless and that new colonies are produced through the fission of existing colonies (in the vast majority of ant species, colonies are begun by individual winged foundress queens, which, theoretically, can raft across water barriers).

The geological data, including the existence of land bridges, suggest that (1) *Aenictus* arose in tropical Laurasia in the early Tertiary and dispersed into Africa sometime between the late Oligocene and the late Pliocene (because *Aenictus* was tropically adapted, dispersal to North America across the North Pacific bridge was not possible); (2) *Dorylus* evolved in Africa in the early Tertiary but did not spread to Asia until late in the Tertiary, before the land connection narrowed and became arid; the lack of diversity in this genus in Asia suggests that dispersal may have occurred even later, during the Quaternary (or competitive exclusion may account for the low number of *Dorylus* species in Asia, since *Aenictus* was well established when *Dorylus* arrived); and (3) the Ecitoninae arose in tropical South America, diversified during a long period of isolation, and did not disperse to North America until the end of the Tertiary (Gotwald 1977, 1979). To summarize: the true army ants appear to be triphyletic, a supposition consistent with the cladistic analyses and conclusions of Bolton (1990c).

Comparative Morphology

Comparative studies of ant morphology are traditionally used to elucidate phyletic relationships. Such investigations have produced specu-
lative accounts of army ant phylogeny. Most authors (see Bolton 1990a)
seem to have regarded ants of the subfamily Cerapachyinae as ancestral
to the army ants, implying, perhaps, that the true army ants are monophyletic. Carlo Emery (1895) expressed his view of this phyletic relationship by placing the cerapachyines as a tribe within the Dorylinae.
Although W. L. Brown, Jr. (1975), the most influential ant taxonomist of
the latter half of the twentieth century, suggested that the ecitonines and
Aenictus may have arisen separately from cerapachyine ancestors, consensus on the relationship between the army ants and this group has
not been achieved. W. M. Wheeler (1928) supposed the true army ants
to be monophyletic and speculated that *Cheliomyrmex* linked the New
and Old World faunas. Schneirla (1971) thought a monophyletic origin
best fitted the revealed functional and behavioral evidence. And George
C. and Jeanette Wheeler (1984), the world's leading authorities on ant
larvae, decided, based on their studies of known larvae, that all army
ants belong in one subfamily. Most recently, Bolton (1990c) grouped the
Cerapachyinae with the true army ants, which, in his opinion, together
form a monophyletic lineage, the doryline complex.

Comparative morphological studies from the past twenty-five years
have produced conflicting results with regard to the phylogeny of the
true army ants. Mouthpart morphology (Gotwald 1969) and the location
and structure of abdominal exocrine glands (Jessen 1987) can be interpreted as supporting a triphyletic origin, whereas the morphology of the
male internal reproductive system (Gotwald and Burdette 1981) and the
ultrastructure of the Dufour's gland lining (Billen 1985, Billen and Gotwald 1988) suggest that *Dorylus* and *Aenictus* are closely related, but-
tressing the diphyletic hypothesis.

A preliminary cladistic analysis of the true army ants, using a limited
number of characters, produced mixed results (Gotwald 1979). All attempts to generate a cladogram that adequately explained the identified
synapomorphies met with failure. The matrix of synapomorphy produced in this study revealed that the fewest character-states are shared
between *Dorylus* and the New World genera *Eciton, Labidus, Neivamyrmex,* and *Nomamyrmex. Cheliomyrmex,* however, shares as many derived
character-states with the New World genera as it does with *Dorylus;* the
same is true of *Aenictus.* Any attempt to construct a cladogram was
confounded by these two genera. Indeed, a phylogenetic tree could not

be constructed unless certain character-states were disposed of as convergently evolved. A similar, but unpublished, cladistic analysis of the subgenera of *Dorylus* that I undertook in 1982 found that the species group together into their respective subgenera, except for *Dorylus (sensu stricto)* and *Anomma*. The latter two share an immediate common ancestor. Ward (1990) and Bolton (1990c) have produced definitive cladistic analyses that may put to rest at least some of the debate on the internal phylogeny of the ants.

An Evolutionary Scenario

It is also possible to speculate about the evolution of behavior in the origin of the army ant adaptive syndrome. Wilson (1958a) formulated and refined (Hölldobler and Wilson 1990) a working hypothesis comprising major adaptive steps in the elaboration of army ant behavior:

1. Group-recruitment raiding develops, permitting specialized feeding on other social insects. That is, group raids are led by scouts to prey discovered by the scouts. This type of raiding apparently exists in some species of the ponerine genera *Leptogenys* and *Megaponera*, among others.

2. Group raids are initiated "autonomously," without the recruiting activities of scouts. These rather more massive raids allow specialized feeding on large arthropods and other social insects and permit a larger part of the trophophoric field to be searched. This type of raiding occurs without frequent migrations and is seen in the myrmicine genus *Pheidologeton*.

3. Nomadism either develops concurrently with group-raiding behavior or is added shortly afterward. Emigrations to new nest sites allow a colony to shift to new trophophoric fields with more abundant prey. This combination of group raiding and nomadism exemplifies army ant behavior and has been achieved in the genus *Leptogenys*.

4. As group predation becomes more efficient, large colony size is possible. This stage has been achieved by numerous species of *Aenictus* and *Neivamyrmex*.

5. Diet may expand to include other smaller and nonsocial arthropods, some small vertebrates, and vegetable matter. Colonies may become quite large. This general predator stage is seen in the driver ants of Africa and in *Eciton burchelli*.

In a similar effort, I categorized the army ants after polarizing their behavioral character-states as either ancestral or derived (Gotwald 1985, 1988). This was done not to construct a series of genealogies but to measure the degree to which any one species may be derived from a hypothetical ancestor. In polarizing the behavioral character-states, I made outgroup comparisons with the Cerapachyinae because this group has so often been considered ancestral to the army ants. I hypothesized that the following character-states are plesiomorphic, or ancestral: (1) hypo-gaeic activity of any kind, (2) column raiding, (3) a nonphasic reproductive cycle (i.e., without alternating nomadic and statary phases), (4) small colony size, (5) specialized predation on other social insects, and (6) a relative lack of caste polyethism. Conversely, apomorphic, or derived, character-states were identified as (1) epigaeic activity of any kind, (2) swarm raiding, (3) a phasic reproductive cycle of alternating nomadic and statary phases, (4) large colony size, (5) expansion of diet to include large and small arthropods in addition to social insects, and (6) a high degree of caste polyethism, implying well-developed polymorphism. The four categories of army ants described below suggest possible stages in army ant evolution, starting with species whose activities are largely subterranean and concluding with those that are almost exclusively surface active.

Subterranean (hypogaeic) species. These nest, forage, and emigrate beneath the soil surface; they are column raiders, specialized predators of social insects, generally form small colonies, and have monomorphic or moderately polymorphic workers. This level of army ant behavior is seen in *Cheliomyrmex* and some species of *Aenictus*.

Presumptively subterranean species. Species in this group are much like the subterranean forms but occasionally forage and emigrate on the substrate surface. *Aenictus asantei*, for instance, is clearly this type (see Campione et al. 1983).

Presumptively surface-adapted species. These nest in the soil but forage and emigrate on the soil's surface; they are swarm raiders, general predators, form large colonies, and have polymorphic workers. The African driver ants have achieved this level (Raignier and van Boven 1955, Gotwald 1974a).

Surface-adapted (epigaeic) species. Similar to the previous category, these species additionally nest on or above the substrate surface; they are most often swarm raiders, are oligophagic or general predators, form large colonies, and have polymorphic workers. Species in this category are the most derived and are exemplified by *Eciton burchelli* (Schneirla 1971).

In commenting on this proposed correlation between the soil zone in which the army ants are active and their degree of prey specialization, Hölldobler and Wilson (1990, p. 595) noted that if the correlation holds, "it is superimposed on the evolutionary grades defined by colony size, group raiding, and nomadism to account for part of the variance in degrees of prey specialization among species."

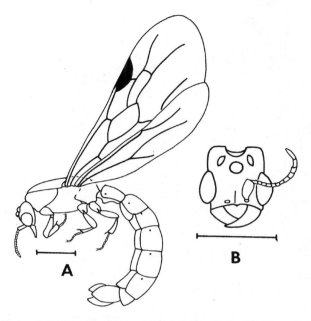

Figure 2.17. Male of *Aenictogiton fossiceps* (A, habitus; B, head). The genus *Aenictogiton*, with its affinities for the Aenictinae and Dorylinae, is known only from males. Each scale equals 1 mm. (Redrawn from Emery 1910.)

Convergent Evolution of the Army Ant Adaptive Syndrome

That the army ant adaptive syndrome is manifestly advantageous in tropical habitats can be demonstrated by noting the number of other ant species that have convergently evolved army ant lifeways. Army ant behavior and concomitant morphological adaptations have been documented, for example, in the ponerine genera *Cerapachys* and *Sphinctomyrmex* (now in the doryline section of the poneroid complex), *Leptogenys, Megaponera, Onychomyrmex, Simopelta,* and *Termitopone* (W. M. Wheeler 1936, Wilson 1958a,b, Gotwald and Brown 1966, Hermann 1968a, Buschinger et al. 1989). The extent to which army ant behavior has arisen in the Ponerinae is best illustrated by *Leptogenys.* One species from Malaysia, *L. distinguenda,* (1) forms temporary nests, or bivouacs (a characteristic of *Eciton*); (2) maintains large colonies, perhaps in excess of 50,000 workers; (3) has a single, physogastric queen (i.e., she has an unusually swollen abdomen) per colony; (4) conducts frequent emigrations to new nest sites; (5) forages en masse (up to 38,000 or more work-

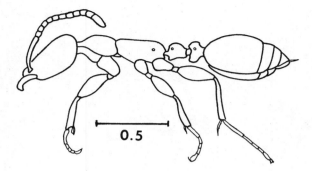

Figure 2.18. Worker of *Leptanilla judaica*. The genus *Leptanilla* comprises miniscule, subterranean ants that behave in some ways like army ants. Scale in millimeters. (Redrawn, by permission of J. Kugler and the Entomological Society of Israel, from J. Kugler 1986.)

ers per raid); and (6) is a general predator, taking as prey a wide variety of adult and immature arthropods (Maschwitz et al. 1989). The way this species combines group predation and nomadism does indeed make it an army ant (Plate 1B).

Most surprisingly, group hunting has evolved in two species of the genus *Pheidologeton*. These species—*P. diversus* and *P. silenus*—are the only members of the subfamily Myrmicinae, the largest of the ant subfamilies, known to hunt cooperatively en masse (Moffett 1984, 1987, 1988a,b). In the field, their foraging behavior is uncanny in its resemblance to that of army ants. I first observed these ants in Malaysia and was startled by their foraging columns, which, seen from a distance, are indistinguishable from those of *Aenictus*. In addition to group hunting, these species have large colonies, comprising perhaps as many as 250,000 workers in *P. diversus* (Moffett 1988b). Otherwise, however, they have not achieved the army ant adaptive syndrome. Colonies are established by foundress queens that initially possess wings, and although *P. diversus* emigrates, it does not do so frequently enough to be considered nomadic (Moffett 1988b).

Two enigmatic groups of ants appear suspiciously like army ants, although little is known about them. One, the tribe Aenictogitini, contains the single genus *Aenictogiton*, which in turn includes but seven species,

Figure 2.19. *Leptanilla japonica*, seen here in laboratory colonies, is the only leptanillinine to have been studied alive for an extended period. Queen, workers, and larvae (A); physogastric queen (B); and workers transporting larvae (C). (Photographs by Keiichi Masuko.)

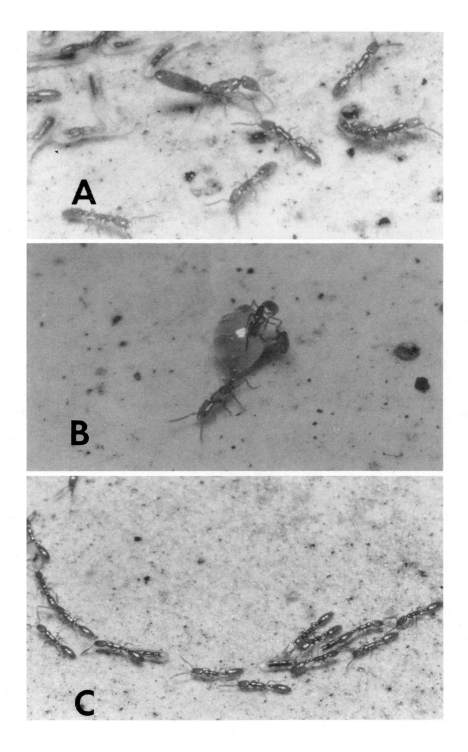

all known only from males. Although the males have an army ant–like habitus, the genus was placed in the Ponerinae (Figure 2.17). Recently the tribe Aenictogitini was elevated to subfamily status and its affinities with the Aenictinae and Dorylinae recognized (Baroni Urbani et al. 1991). The genus seems confined to the Congo Basin (W. L. Brown 1975).

The second group is the intriguing subfamily Leptanillinae. With fewer than 50 species, many known only from males (Kutter 1948, Baroni Urbani 1977, J. Kugler 1986), the subfamily has seldom been observed alive for extended periods (Figure 2.18)—with one notable exception. Keiichi Masuko (1987) collected 11 colonies of the cryptic species *Leptanilla japonica* and observed them in laboratory nests. His observations confirm, to a large extent, the traditional view that the leptanillinines are army ants. These miniscule ants are strictly subterranean; each colony has but a single queen; the queens are typical dichthadiigynes; brood development is synchronized and cyclical; the workers forage over trunk trails; when laboratory nests were disturbed, the workers immediately started to transport their larvae along the trail system (demonstrating a propensity to emigrate); and the species appears to be a specialized predator of geophilomorph centipedes (Masuko 1989, 1990) (Figure 2.19).

Hölldobler et al. (1989) additionally discovered that the *L. japonica* queen is well endowed with exocrine glands, a characteristic shared by other army ant queens. The adults are unusual in that they are able to feed on hemolymph taken from their own larvae. They imbibe the hemolymph from the larval body cavity through specialized duct organs, one on each side of the fourth abdominal segment, referred to as "larval hemolymph taps." Larval hemolymph is the sole source of nutrients for the *L. japonica* queen (G. C. Wheeler and Wheeler 1988; Masuko 1989, 1990). This feeding mechanism is decidedly unlike that of army ants.

There can be little doubt that the advantage of being an army ant lies in the quantitative and qualitative expansion of the diet that the lifeway permits. Certainly, army ants have access to a wide range of prey not available to solitary foragers, and apparently this is a significant adaptive advantage in tropical habitats. The army ant adaptive syndrome is an evolutionary attempt at predatory perfection.

3

The Colony

Designers and architects who embrace the axiom "Form follows function" believe that the shape of a utilitarian object should be dictated by its use. Philosophically, they object to the antique coffee grinder that is adroitly converted into a table lamp. They would, on the other hand, feel quite comfortable with the utility of design chosen by natural selection. Biologists who specialize in the morphology and anatomy of organisms are acutely aware of nature's adherence to this dictum on form.

It should not surprise us that an organism's fitness within a particular set of environmental circumstances depends on the adaptiveness or serviceability of its design, whether it is the configuration of a hemoglobin molecule or the surface area of a pectoral fin. But nature's designs are characterized by their heterogeneity, and this variation in design within a population of organisms is the clay with which natural selection sculpts evolving designs in the ever-changing struggle for existence. It is imperative, therefore, to investigate and understand the designs of organisms and their relation to adaptive and evolutionary success. Knowledge of designs or form, as we shall see, is prerequisite to comprehending the functioning of colonial life.

The ecological success of insects rests, to a great extent, on three singular events in their evolution: the acquisition of wings and flight, the appearance of metamorphosis, and the achievement of colonial life (Wilson 1985a). In turn, the competitive advantage of colonial (i.e., social) insects, as evidenced in their prodigious numbers, is at least partly attributable to the existence of adult specialists, or castes, that carry out particular tasks. It is exceedingly important, then, to identify the ways by which colony members differentiate into castes and partition labor (Wilson 1985a). Most often a caste is a morphological entity whose size and proportions facilitate its tasks. In other words, form follows function. The coexistence of two or more castes within the same sex is called polymorphism (Wilson 1971a).

The nature of colonial life as it relates to caste production involves ontogeny—the course of development of the individual—and a process at the colony level called sociogenesis (or colony ontogeny), defined as "the procedures by which individuals undergo changes in caste, behavior, and physical location incident to colonial development" (Wilson 1985a, p. 1489).

How do different colony members develop into different castes? It appears that the developmental origin of such differences is primarily, if not exclusively, environmentally induced rather than genetically fixed. To wit, all females have essentially the same genetic potential to become any caste, but environmental factors are the final determinants. These factors include larval nutrition, inhibition caused by pheromones, egg size, winter chilling, temperature during development, and age of the queen (Wilson 1985a).

Colony organization in army ants, especially as it relates to caste and caste-correlated division of labor, is the focus of this chapter.

Morphology, Ontogeny, and Polymorphism

"The colony can be most effectively analyzed if it is treated as a factory within a fortress," Hölldobler and Wilson (1990, p. 298) observed. In their analogy, the factory consists of the workers, which gather energy that is converted into new reproductives (i.e., queens and males) and ultimately into new colonies. The fortress is the defense system—an array of stings, noxious secretions, and lethal mandibles—that the colony can muster against predators. The ant colony is a bonanza of protein, fat, and carbohydrate, a ready-made banquet for the ravenous predator. Ultimately, the successful operation of this factory within a fortress depends on the adaptive arrangement of castes and the division of labor among these castes.

The meaning of the term *caste*, as used in the literature, is rather elusive. Commonly, all workers, as distinct from the queen, are referred to as a single caste, and morphologically and behaviorally distinct subsets of workers are called subcastes. Not infrequently, however, the subcastes are alluded to as castes in their own right. Regardless of what they are called, the identification of these specialist groups is important to our understanding of colony life. Such groups can be identified as physical castes (or subcastes), distinguished by their morphology; temporal castes (or subcastes), defined by their age; or physiological castes (or subcastes), characterized by their paramount physiological state (Hölldobler and Wilson 1990). Thus, even though a colony member may belong to one physical caste, as it ages it may progress through a sequence of work roles or temporal castes (Wilson 1985a).

The morphological characteristics that distinguish one female caste from another are most often attributable to allometry, differential growth of body parts in relation to total body size. That is to say, allometric growth occurs when two different body parts grow at different exponential rates (Wilson 1953). Take, for instance, the relationship between head capsule size and mandible morphology in the New World army ant *Cheliomyrmex morosus* (Gotwald and Kupiec 1975). The smallest workers of this species possess a somewhat triangular mandible with the apical and subapical teeth separated by a series of denticles, whereas the largest workers have a falcate mandible with a sharply pointed apical tooth and two subapical teeth (Figure 3.1).

Allometric growth in polymorphic species thus produces mandibles adapted to perform different tasks (Gotwald 1978b). This adaptiveness in mandible design did not escape the scrutiny of the nineteenth-century scientific naturalist Henry Walter Bates. A contemporary and colleague of Charles Darwin and Alfred Russel Wallace, Bates spent eleven years in the Amazon Basin observing and collecting its incredibly diverse biota. He was especially fascinated with mandible morphology in *Eciton* (Bates 1863, p. 368):

> Like many other ants, the communities of Ecitons are composed, besides males and females, of two classes of workers, a large-headed (worker-major) and a small-headed (worker-minor) class; the large-heads have, in some species, greatly lengthened jaws, the small-heads have jaws always of the ordinary shape; but the two classes are not sharply-defined in structure and function, except in two of the species. There is in all of them a little difference amongst the workers regarding the size of the head; but in some species this is not sufficient to cause a separation into classes, with division of labour; in others the jaws are so monstrously lengthened in the worker-majors, that they are incapacitated from taking part in the labours which the worker-minors perform; and again, in others the difference is so great that the distinc-

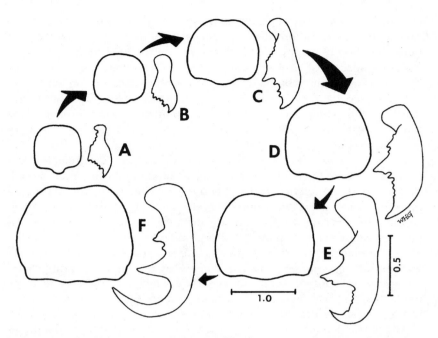

Figure 3.1. Head capsules and respective left mandibles of *Cheliomyrmex morosus* workers, selected to demonstrate the nature and range of polymorphism and allometry in this and other army ant species. Allometric growth in polymorphic species produces mandible morphologies that are variously adapted to different tasks. Scales in millimeters. (Redrawn, by permission of the Entomological Society of America, from Gotwald and Kupiec 1975.)

tion of classes becomes complete, one acting the part of soldiers, and the other that of workers.

Even the illustrious Charles Darwin (1872, p. 206) was fascinated with army ants, not so much for their reputation as ferocious foragers as for their remarkable display of allometrically derived polymorphism. After examining a series of workers of the subgenus *Anomma*, he drew an interesting analogy about their morphology: "The difference [between worker ant sizes] was the same as if we were to see a set of workmen building a house, of whom many were five feet four inches high, and many sixteen feet high; but we must in addition suppose that the larger workmen had heads four instead of three times as big as those of the smaller men, and jaws nearly five times as big."

With the exception of the genus *Aenictus* and some species of the genus *Neivamyrmex*, all true army ant workers, both ecitonines and dorylines, are strongly polymorphic and therefore exhibit a wide range of allometrically produced morphological differences. Most apparent on cursory examination are the differences in total body length.

The size differential between the smallest and largest workers in *Eciton burchelli* is 8.1 mm; in *Dorylus* (*Anomma*) *wilverthi*, 8.0 mm; in *Cheliomyrmex morosus*, 4.12 mm; and in *Neivamyrmex nigrescens*, 2.8 mm; but in *Aenictus gracilis*, a monomorphic species, it is 0.5 mm (Gotwald and Kupiec 1975). Furthermore, the size frequency distribution of workers of polymorphic species is asymmetric. For example, in *E. hamatum*, small intermediate workers are most numerous, large intermediates and minors are next, and major workers are least numerous (Schneirla 1971). In *E. burchelli*, the minor workers predominate and major workers constitute less than 2 percent of the entire colony (Schneirla 1971), although Franks (1985) found that the caste ratios of callow workers (i.e., those recently emerged from the pupal stage) change as a function of colony size. Topoff (1971), who calculated the frequency distributions of total body length for army ant worker pupae, noted that the smaller workers predominate even in essentially monomorphic species such as *A. laeviceps*.

Allometric growth was first analyzed in army ants by J. S. Huxley (1927), who called it heterogonic growth. Huxley quantitatively examined allometry in the Old World genus *Dorylus*, subgenus *Anomma*. Although François Cohic (1948) claimed that *Anomma* workers could be divided into distinct morphological and functional types, M. J. Hollingsworth (1960) declared that such discontinuities did not exist and demonstrated that *Anomma* workers of a single colony could be placed in a continuous series from smallest to largest. On the other hand, van Boven (1961) found four morphologically distinguishable subcastes in *D.* (*A.*) *wilverthi*, which he called minima, minor, media, and major workers. Hollingsworth (1960) noted that allometry in *Anomma* is not simple, since different parts of the worker's body have different allometric constants in different individuals. The relative growth curve of the workers he measured changed slope, a condition referred to as diphasic allometry. This circumstance is identifiable when an allometric regression line plotted on a double logarithmic scale separates into two sections of different slopes that meet at an intermediate point (Hölldobler and Wilson 1990). Franks (1985) reported a quadrimodal distribution of *E. burchelli* worker sizes (Figure 3.2). Although this size frequency distribution may be statistically evident, it is not readily obvious in a cursory examination of workers under the microscope.

Differential growth that results in polymorphism among adult workers begins during larval development when imaginal discs grow at dif-

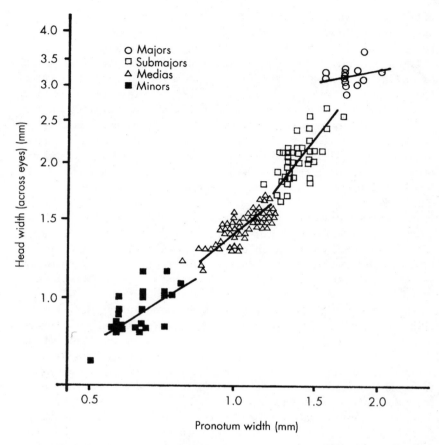

Figure 3.2. Head width–pronotum allometry in *Eciton burchelli* workers, in which four distinct castes can be recognized. Curves have been fitted to each of the four castes by the least squares method. (Redrawn, by permission of N. R. Franks and Gustav Fischer Verlag, from Franks 1985.)

ferent rates. These discs are composed of undifferentiated cells that give rise to adult organs during pupal development (Hölldobler and Wilson 1990). Measurements of body length and imaginal disc areas and the relationship of these values in brood samples from *Eciton* and *Neivamyrmex* have demonstrated rather conclusively that brood allometry is regular throughout much of the larval stage. Although the first eggs to hatch remain the most developmentally advanced during most of larval life, and the last to hatch remain the least developed, this time range difference is shortened considerably by differential growth patterns in the larvae and pupae (Tafuri 1955, Lappano 1958). That is, the youngest larvae catch up with the older larvae and emerge from the pupa at approximately the same time, but as smaller adults with different body proportions. Schneirla et al. (1968, p. 552) called this shortening of the egg-hatching time differences developmental convergence.

Castes and Subcastes

Workers

Morphology

Almost all army ant species lead a primarily subterranean existence and exhibit a concomitant reduction in visual acuity. The workers of *Dorylus*, *Aenictus*, and *Cheliomyrmex* are eyeless, while, with a few exceptions, those of *Eciton*, *Labidus*, *Neivamyrmex*, and *Nomamyrmex* possess compound eyes reduced to a single facet. Even blind species remain sensitive to light intensity, however, indicating that they may possess a subdermal or integumental light sense (Werringloer 1932). Although several species of *Dorylus*, especially the driver ants, and *Aenictus*, *Eciton*, and *Neivamyrmex* are adapted to life on the surface in most aspects of their biology, their reduced eyes suggest a relatively recent and secondary return to activity aboveground. The reduced palpal segmentation in the mouthparts of army ants supports this hypothesis of subterranean existence for the immediate ancestors of the surface-adapted species (Gotwald 1978b).

Celebrated for their ability to pierce human skin, the falcate, sharply pointed mandibles of the soldier subcaste of the army ants *Eciton* and *Dorylus* (*Anomma*) and the major workers of *Cheliomyrmex* are nevertheless the exception in the diverse mandible morphology of the army ants.

The mouthparts of army ant workers are morphologically unique at the tribal level, and the Ecitonini, Aenictini, and Dorylini can be distinguished from one another on the basis of mouthparts alone (Figure 3.3). The number of palpal segments in army ants is significantly reduced from the primitive number in ants: from six in the maxillary palpus to one or two segments, and from four in the labial palpus to two or three segments (Gotwald 1969).

The middle body region, which bears the legs, is the alitrunk or mesosoma; as in all other apocritan Hymenoptera, it consists of the three thoracic segments plus the first abdominal segment. The latter segment is so completely fused to the third thoracic segment that it is essentially indistinguishable from it. Two distinct types of mesosoma can be found in the army ants. One, typical of the Dorylini, consists of two parts of

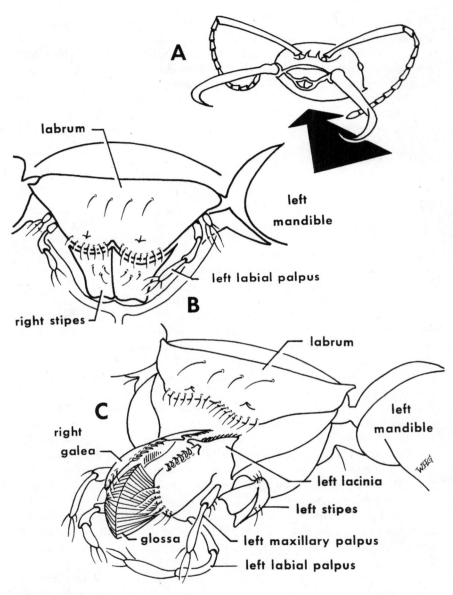

Figure 3.3. Mouthpart morphology of the *Eciton mexicanum* major worker, or soldier. Anterior view of the head, mandibles spread and maxillo-labial complex retracted (A) (arrow indicates direction of view for drawings B and C); the mouthparts with the maxillo-labial complex retracted (B) and extended (C). (Redrawn, by permission of Cornell University, Agricultural Experiment Station, Geneva, from Gotwald 1969.)

approximately equal size; the other, representative of the Ecitonini and Aenictini, is an undivided, carapace-like structure in which suturing is greatly reduced (Reid 1941) (see Figures 2.5–2.7, 2.10, 2.11, 2.13, 2.15). The worker's waist, which connects the mesosoma to the gaster, is bi-nodal (two-segmented) in the Ecitonini and Aenictine and uninodal (single-segmented) in the Cheliomyrmecini and Dorylini. That both conditions exist in the Ecitoninae suggests that segmentation of the waist is far less conservative in the New World army ants than in other ant subfamilies (or that they are not as closely related as the classification portrays them to be).

B. E. Pullen (1963) suggested that the binodal waist facilitates stinging by making the gaster more maneuverable, and Schneirla (1971) added that this condition helps surface-adapted species to subdue strong, fast-moving prey. Schneirla also concluded that the flexibility is of some advantage in laying chemical trails and in carrying brood and prey, which are slung beneath the worker's body.

The worker's gaster (the abdomen minus the first two or three segments, depending on whether the waist is uninodal or binodal) is relatively undistinguished; it terminates in the sting, which, as in all aculeate Hymenoptera, is derived from the ovipositor. In *Dorylus*, the gaster is exceptional in that the pygidium, or terminal tergite (dorsal sclerite), is impressed with a circular concavity. The pygidium is armed on each side of this concavity with a single pygidial spine. Distally each spine terminates in one, two, or three apices (Gotwald and Schaefer 1982). The functional significance of the pygidial concavity and spines constitutes a provocative enigma.

Among the complex of sclerites composing the sting apparatus, two are of special interest. The furcula, a sclerite located anterior to the sting bulb, is common to all ants thus far examined except the Aenictinae, Dorylinae, Ecitoninae, Cerapachyinae, and the ponerine species *Simopelta oculata* (Hermann 1969). Its absence in the army ants and in *S. oculata*, an ant with army ant lifeways, suggests that this sclerite is convergently lost in species that evolve the army ant adaptive syndrome (Hermann 1968b). H. R. Hermann and J. T. Chao (1983) concluded that the lack of a furcula lessens sting manipulation but that this may be relatively unimportant in army ants, which subdue their prey primarily through biting and tearing rather than by stinging.

The sting itself, the other sclerite of interest, is broad and spatulate in the Dorylini, a morphological development correlated with this group's apparent inability to sting. This sclerite is slender in the army ant species that do sting (Hermann 1969, Gotwald 1978b). All New World army ants possess a functional sting, and when attacking prey these species

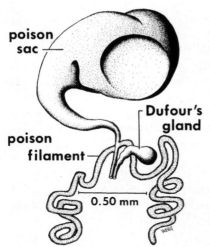

Figure 3.4. Glandular elements of the poison apparatus of the *Cheliomyrmex morosus* soldier. (Reprinted, by permission of the New York Entomological Society, from Gotwald 1971.)

may bite and sting simultaneously. Species of the Old World genus *Dorylus*, however, are not known to sting, although they bite with consummate skill. The *Aenictus* workers Schneirla (1971) observed in the Philippines had potent stings, but the African *Aenictus* observed by Gotwald (1978b) did not sting.

The relative importance of the sting to the evolution of eusociality in the aculeate Hymenoptera—or of eusociality to the origin of the sting—remains unresolved in the face of continued debate (see Starr 1985, Kukuk et al. 1989). C. K. Starr (1985) argued that the sting appeared early in the Hymenoptera and enabled incipient social evolution by permitting groups to defend themselves against vertebrate predators. The survival of such groups, he suggested, ensured the continued evolutionary progress of sociality. The sting, countered P. F. Kukuk et al. (1989), was not a significant preadaptation in facilitating—indeed was not even necessary in—the origin of group living and the early evolution of eusociality.

Soft parts of the poison apparatus associated with the sting include an elongate, pear-shaped or spherical poison sac or venom reservoir with a conspicuous duct that terminates in the sting bulb (Hermann 1969, Gotwald 1971). Free poison filaments, sometimes branched, attach to the base of the poison sac or its duct (Figure 3.4).

The poison apparatus of *Eciton burchelli* produces venom that exhibits phospholipase, hyaluronidase, protease, and acid phosphatase activity (Schmidt et al. 1986). Interestingly, *E. burchelli* venom has the highest

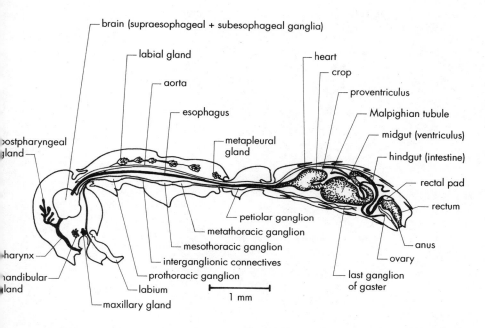

Figure 3.5. Major features of the internal morphology of the *Dorylus* (*Anomma*) *nigricans* worker (poison apparatus and gaster glands omitted). (Based, with permission, on an original drawing by J. K. A. van Boven.)

protease levels of any insect venom analyzed. Among the ants thus far examined, this species has by far the highest acid phosphatase activity (1000 times above the detection limit) (Schmidt et al. 1986), although no role for the elevated level of this common tissue enzyme has been identified. Acid phosphatase is prevalent in many ant species and has been reported as well from honey bees, bumblebees, and yellowjackets. The role of protease remains a mystery also. Army ants gather large amounts of arthropod prey that must be consumed before it decomposes, and J. O. Schmidt et al. (1986) thought the idea that the protease might predigest prey and thereby accelerate digestion was intuitively attractive. That is, prey tissues might begin to break down while the prey were still in the clutches of their homeward-bound captors. Rettenmeyer, however, reported to Schmidt and his colleagues that *Eciton* does not normally use its sting in prey collecting.

The internal morphology of army ant workers is incompletely known

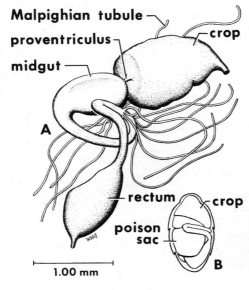

Figure 3.6. Digestive system of the *Cheliomyrmex morosus* soldier, as it appears when removed from the gaster (A) and in place within the gaster (B). (Reprinted, by permission of the New York Entomological Society, from Gotwald 1971.)

(Figure 3.5). The digestive tract is relatively simple (Figure 3.6). The pharynx of *Anomma,* for instance, is reduced, and this reduction may be correlated with the relative absence of worker food exchange, or trophallaxis, in army ants. The proventriculus, a valve that closes the crop, or social stomach (a storage compartment from which liquid food can be regurgitated to nestmates), is degenerate in army ants (Figure 3.6). Since the damming action of this membranous valve probably depends on muscle contraction, crop storage in army ants may be of short duration (Eisner 1957).

The Malpighian tubules, excretory structures that insert near the junction of the midgut and hindgut, are probably histologically uniform throughout the army ants, although the number of tubules varies considerably from species to species (Figure 3.6). Even so, overlap in the ranges of tubule numbers between species is common, and the number of tubules per individual is so closely correlated with body size as to preclude their use taxonomically in polymorphic species (Gotwald 1971). The rectal papillae or pads—structures important in the reabsorption of water, salts, and amino acids in insects—are more constant in number, although R. M. Whelden (1963) reported a range of three to six in *E. burchelli* and *E. hamatum.*

Like other formicids, army ants are crammed full of glands and sundry secretory cells. Moving about the environment they seem no less than mobile chemical factories. Glands thus far identified in the workers are the mandibular glands, maxillary glands, pharyngeal glands, labial glands, metapleural glands, Dufour's gland, the convoluted gland of the poison sac (Gotwald and Kupiec 1975), pygidial glands, postpygidial glands, and a glandular epithelium in the seventh abdominal sternum (Hölldobler and Engel 1978).

The paired metapleural glands of the posterior thorax, unique to the ants and once considered universally present throughout the Formicidae, are now known to be absent in numerous *Camponotus* species and other formicines (Figures 3.7, 3.8) (Hölldobler and Engel-Siegel 1984). Metapleural secretions have a demonstrated antiseptic effect and may protect the ant's body surface and the nest against invading microorganisms (Maschwitz et al. 1970). These secretions have been shown, for instance, to suppress the germination of spores of various soil fungi (Beattie et al. 1986). In *Dorylus* each gland consists of 40 to 50 spherical cells (Billen and van Boven 1987); in *E. hamatum* soldiers, 180 cells; and in *Neivamyrmex nigrescens,* 201 cells (Hölldobler and Engel-Siegel 1984).

The Dufour's gland, an elongated sac whose wall consists of a monolayered epithelium, opens through the sting in ants. Curiously, the lining of this gland is smooth in all ants except the army ant genera *Dorylus* and *Aenictus,* in which the lining is crenellate (Figure 3.9) (Billen 1985, 1987; Billen and Gotwald 1988). The contents of the Dufour's gland of *D. (A.) molestus* and *D. (A.) nigricans* consist chiefly of linear alkenes and alkanes, primarily tricosene and tricosane (Bagneres et al. 1991). In *molesta* these two compounds constitute 70 percent of the total. In this, the first such analysis of the contents of any gland in the Dorylinae, Anne-Geneviene Bagneres et al. (1991) noted that the principal components have a relatively high molecular mass (C_{23} and C_{25}), which is characteristic of tropical species (in temperate ant species, C_{11}–C_{17} compounds predominate). Small amounts of dihydrofarnesol were also detected.

In a subsequent study, the Dufour's gland of *Eciton burchelli* workers was found to contain a mixture of higher alkanes, alkenes, methyl-branched alkanes, and some oxygenated and terpenoid compounds. The secretion of the *E. burchelli* soldier is more variable than that of the worker, sometimes including ocimene and geranylacetone or large amounts of geranyllinalool (Keegans et al. 1993). The Dufour's gland chemistry of the workers is like that of many other ants. The Dufour's gland of *Labidus praedator* and *L. coecus* are notable for containing as a major substance a relatively volatile monoterpene, (*E*)-β-ocimene—the first monoterpene to be found as a major substance of the Dufour's gland

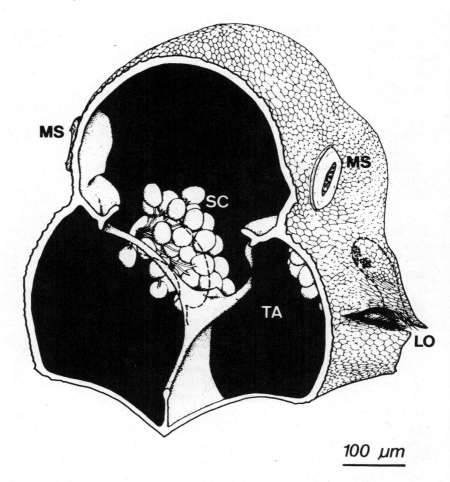

Figure 3.7. Posterior part of the mesosoma of a *Dorylus* (*Anomma*) *molestus* worker, showing the general organization of the metapleural gland. LO, slitlike metapleural gland opening; MS, metathoracic spiracles; SC, secretory cells; TA, thoracic apophyses or supports. The dotted line outlines the metapleural gland's reservoir-like atrium. (Reprinted, by permission of J. Billen and the African Museum, Tervuren, from Billen and van Boven 1987.)

of an ant (Keegans et al. 1993). Ocimene has a pleasant odor for humans, was first isolated from the leaves of the herb sweet basil, and is a constituent of many perfumes.

Little is known about the morphology or developmental anatomy of the army ant nervous system, although the histology of the system has been examined in *Cheliomyrmex morosus* (Gotwald and Kupiec 1975), *E.*

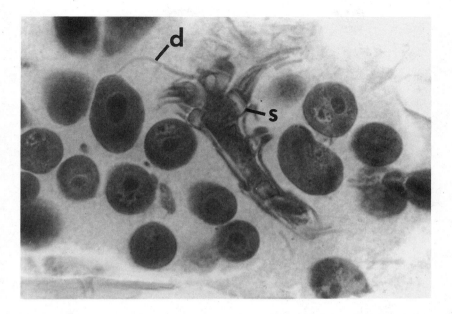

Figure 3.8. The metapleural gland of a *Cheliomyrmex morosus* worker. The secretory cells are conspicuous in this histological cross section through the posterior portion of the gland (400×). Each cell gives rise to a small duct (d) of intracellular origin. Cell products are secreted through these ducts, which unite to form bundles, onto sieve plates (s) located at the apex of an accessory metapleural structure. (Reprinted, by permission of the Entomological Society of America, from Gotwald and Kupiec 1975.)

burchelli and *E. hamatum* (Whelden 1963), *Aenictus asantei* (Campione et al. 1983), and *D.* (*A.*) *molestus* and *D.* (*A.*) *nigricans* (Gotwald and Schaefer 1982). The brain consists of a fused mass of nervous tissue composed of the supraesophageal and subesophageal ganglia. This mass of nervous tissue is perforated by the esophagus and its associated musculature (the circumesophageal connectives are obscured).

In *A. asantei* and other species, as in insects in general, three paired regions can be identified in the brain: the protocerebrum, the deutocerebrum, and the tritocerebrum. Within the protocerebrum, the largest of the three, the corpora pedunculata, or "mushroom bodies," are clearly delineated. The relative size of the corpora pedunculata in the ant brain has been used as an indicator of "mental capacity"; it is smaller in *E. hamatum* compared with non–army ants (Vowles 1955, Bernstein and Bernstein 1969).

The relative size of the army ant brain is predictably smaller than in

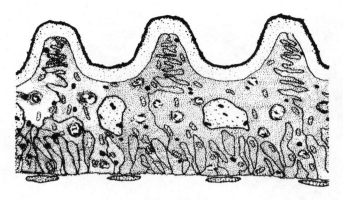

DORYLINAE

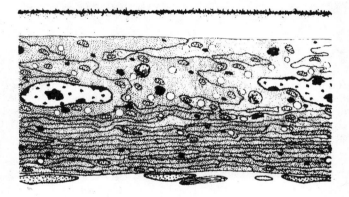

ECITONINAE

Figure 3.9. The Dufour's gland epithelium in the Dorylinae (and Aenictinae), in which it is crenellate, and the Ecitoninae, in which it is not. (Reprinted, by permission of J. Billen and the Union Internationale pour l'Etude des Insectes Sociaux, Section Française, from Billen 1985.)

other similarly sized ants because the optic centers are greatly, and understandably, reduced (Werringloer 1932, Vowles 1955). The deutocerebrum is characterized by the antennal lobes; the tritocerebrum is retained as a pair of small bodies concealed under the olfactory, or antennal, lobes, innervating the labrum and the walls of the pharynx (W.

M. Wheeler 1910). Because they are eyeless, *Dorylus* and *Aenictus* do not possess the conspicuous optic nerves found in *Eciton*.

In addition to the brain, the central nervous system is completed, as in other insects, by a ventral nerve cord consisting of a series of ganglia from which numerous nerves emanate. The ganglia are joined by paired interganglionic connectives. Three ganglia are found in the mesosoma, one in the petiole, one in the postpetiole (when this segment is present; that is, when the waist is binodal), and one to four in the gaster. Excluding the subesophageal ganglion, there are six ventral ganglia in *Cheliomyrmex* and *Eciton* and eight in *Dorylus* and *Aenictus* (Campione et al. 1983). The number of ganglia in the gaster depends on the extent to which ganglionic fusion has occurred.

Although worker ants are a nonreproductive laboring caste, they commonly possess ovaries. Often these produce trophic eggs, which are fed to the queen or larvae (Hölldobler and Wilson 1990). Army ants are no exception. Their ovaries are paired, and each consists of one to three polytrophic ovarioles, which contain developing oocytes separated by groups of trophocytes, or nurse cells, from which the oocytes receive nutrients (Figure 3.10). Ovaries have been found in the workers of *Dorylus*, *Eciton*, and *Cheliomyrmex* (Mukerjee 1933, Whelden 1963, Gotwald 1971, Gotwald and Schaefer 1982).

Studies of cell ultrastructure (e.g., Billen 1985) and chromosomes in army ants are few indeed. Karyotypes in army ants, for example, have been established for only three species: *Aenictus brevicornis*, 2n = 24; *A. sp.* (near *camposi*), 2n = 30; and *A. laeviceps* (?), 2n = 22 (Hung et al. 1972, Imai et al. 1984a,b).

Worker Functions

"All Nature seems at work," wrote Romantic poet Samuel Taylor Coleridge. For none of earth's creatures does this appear more accurate than for the army ant worker. Her function is labor—at times strenuous, agonistic, and life-threatening, but not infrequently quiescent and unruffled. Of course, worker polymorphism, which is present to some degree in all army ants except those of the genus *Aenictus*, is a special adaptive characteristic. It creates morphological specialists that assume specific duties within the colony (Wilson 1953). Thus, with the exception of *Aenictus*, we would expect a caste- or subcaste-correlated division of labor in army ants. Howard Topoff (1971) noted that workers of different sizes in *Eciton*, *Labidus*, and *Neivamyrmex* exhibit size-correlated differences in sensory thresholds to chemical stimuli, whereas workers of *Aenictus* all react similarly to, for instance, arousal stimuli. Although it has

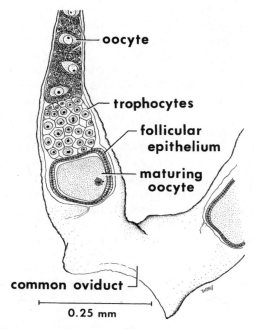

oocyte

trophocytes

follicular epithelium

maturing oocyte

common oviduct

├──────────┤
0.25 mm

Figure 3.10. Right ovary of a *Cheliomyrmex morosus* soldier. The ovary consists of a single polytrophic ovariole containing developing oocytes separated by groups of trophocytes from which the oocytes receive nutrients. (Reprinted, by permission of the New York Entomological Society, from Gotwald 1971.)

not been clearly documented in army ants, we would expect as well that an individual's role in the colony might change with advancing age.

Caste polyethism was most elegantly demonstrated among the New World army ants by Franks (1985), who identified four distinct castes in *E. burchelli:* minims, medium workers, submajors, and majors. Of particular interest are the submajors, which specialize in prey transport (Franks 1985). Although members of this caste or subcaste are only 3 percent of the worker population, they represent 26 percent of the ants carrying prey. The submajors therefore constitute a porter caste. They are the largest *E. burchelli* workers to carry brood and prey; the majors, or soldiers, never do, perhaps as a consequence of their exaggerated hooklike mandibles, which appear poorly designed as grasping structures (Topoff 1971). *E. burchelli* workers frequently form teams that cooperate in transporting large prey items. Significantly, there is a tendency for each group to have a single submajor member (Franks 1986), which led Franks (1986, p. 428) to suggest that "[transport] groups have

a definite sociological composition." What is it about submajor morphology that best adapts this caste to transporting prey? Perhaps most
important, submajors have the longest legs in proportion to body size
of all *E. burchelli* workers. The ratio of hind leg length to body length is
less than 1 for minims, between 1 and 1.3 for medium workers, between
1.4 and 1.5 for submajors, and between 1.3 and 1.4 for majors (Franks
1985). Although burdened ants of all participating castes run at the same
speed, suggesting that leg size does not contribute to retrieval speed, the
submajors carry disproportionately larger prey items (Franks 1985).
Long legs may make such sizable burdens possible, since prey, like
brood, are carried by the workers beneath their bodies.

In addition to the submajors, two of the remaining three castes of *E.
burchelli* are also committed to specialized tasks (Franks 1985). The medium workers are generalists that appear to be involved in all roles
within the colony. The minims, Franks (1985) pointed out, are most common in the bivouac, where they are thought to serve as brood nurses
(Schneirla 1971). In *E. hamatum* and *Neivamyrmex nigrescens*, the minims
also appear to specialize in handling the eggs and newly hatched larvae
and in feeding the smallest larvae (Topoff 1971). The majors, with their
stunning mandibles, specialize in defending the others, possibly even
against potential vertebrate predators (Schneirla 1971). The defensive
function of major workers in *Eciton* becomes evident when a bivouac is
torn apart. In response to this perturbation, the major workers gather
around the queen, a role usually assumed by the smallest workers in
the undisturbed bivouac (Rettenmeyer 1963b). Rettenmeyer (1963b)
found that the major workers of *Eciton* also participate in capturing prey
and in the formation of the hanging clusters essential to assembling the
bivouac. In *Labidus coecus*, the largest workers are the first to arrive at
sites of intense stimulation during raiding (Rettenmeyer 1963b), and in
N. nigrescens, major workers flank the emigration trail at certain times
during the nomadic phase, exhibiting an alarm reaction that consists of
rearing up on their hind legs, vibrating their antennae, and opening and
closing their mandibles (Topoff 1971).

Among the Old World driver ants of the genus *Dorylus*, the size of
the workers is correlated with the tasks they perform in the retrieval of
prey. Most foragers return to the nest without visible prey, a phenomenon that characterizes *Eciton* as well and that led Schneirla (1971) to
conclude that raiding for prey was not particularly efficient. Albert Raignier and van Boven (1955) estimated that 6 to 22 percent of the returning
foragers of *Dorylus* (*Anomma*) *wilverthi* carry prey and only 0.8 to 10
percent of the returning workers of *D.* (*A.*) *nigricans* do so (Plate 2A).
But my own dissection of the returning workers of *D.* (*A.*) *molestus* re-

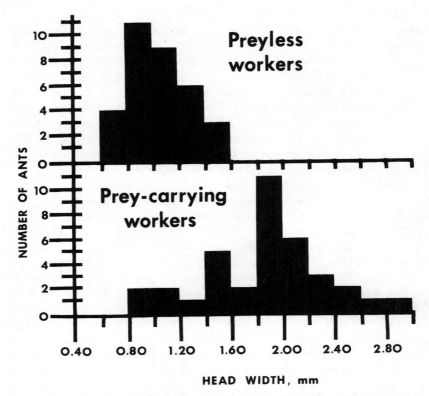

Figure 3.11. Frequency distributions of head widths of preyless and prey-carrying workers of *Dorylus* (*Anomma*) *molesta*. Prey retrieval tasks are correlated with worker size. (Reprinted, by permission of the Entomological Society of America, from Gotwald 1974a.)

vealed that their crops contained a clear, yellowish, viscous liquid, presumably of prey origin; workers going out to forage invariably had empty crops (Gotwald 1974a). Measurement of prey-carrying and preyless workers showed that larger workers carry prey items and constitute a porter caste like that of *E. burchelli*, while smaller workers carry prey fluids (Figure 3.11). In other caste-correlated tasks in *Dorylus*, soldiers assume a defensive posture at the borders of foraging and emigration columns not unlike the behavior of *Eciton* majors (Plate 2B); small workers construct the soil particle walls that commonly border both foraging and emigration trails (Gotwald 1982).

A caste's value, Franks (1985, p. 104) noted, is determined both by its current performance and by the costs of producing and maintaining it and the rate at which its members need to be replaced. Although larger workers are more expensive for the colony to grow, the cost of maintaining them may not be much greater than for smaller workers (Franks 1985). Franks (1985) estimated that the minimum average longevity for an *E. burchelli* worker had to be at least 280 days in order for a colony to be able to grow at all. He speculated further, in the absence of firm data on caste-specific death rates, that submajors and majors live longer than their smaller sister workers. If this is true, then the relative costs of these workers to the colony will be reduced. Watkins and Rettenmeyer (1967) determined that army ant workers live longer in the presence of their queen, most likely because of certain secretions they lick from the queen's body.

But longevity is not solely a function of normal biological senescence. Being an army ant is a risky occupation. Army ant workers probably suffer considerable losses when foraging, especially when the prey are well equipped to defend themselves. Injured stragglers return from raiding forays long after their sister foragers have reached the nest. Their return is fraught with danger as, wounded and battered, many fall prey to other ants—opportunists, especially of the subfamily Myrmicinae and the formicine genus *Oecophylla*, that easily overpower them (Plate 3A) (Gotwald 1982). Schneirla (1971) presumed that swarm raiders must lose many workers to the toxic defensive secretions of prey. Furthermore, he suggested, subterranean foragers must suffer high mortality from the secretions of termites with nasute soldiers, a caste whose head is drawn out into a conical organ that fires, like a cannon, a viscous, sticky substance that entangles the target insect.

Environmental factors may also play a role in longevity. Increasing atmospheric dryness, for example, may take its toll, particularly among the smallest workers, which appear to be the most susceptible to the fatal consequences of desiccation (Schneirla et al. 1954). Their vulnerability to low humidity could be anticipated, since smaller animals have a greater surface area per given unit of volume than do larger animals. A number of factors play a role in workers' resistance to desiccation, not the least of which are cuticular lipids. Ground-dwelling ants (presumably including army ants) are less resistant to desiccation than are arboreal species because they are less amply endowed with these lipids (Hood and Tschinkel 1990). Many workers probably succumb to excessive exposure to solar heat, and certainly others must drown when caught in the torrential rains typical of tropical moist forests.

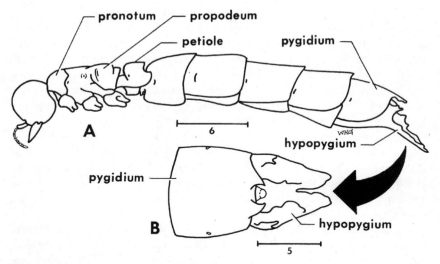

Figure 3.12. External features of the *Dorylus* (*Anomma*) *molestus* queen. Lateral view (A), legs omitted; bifurcated hypopygium in dorsal view (B), showing the uniquely developed last ventral sclerite of doryline queens. Scales in millimeters. (Redrawn, by permission of Blackwell Scientific Publications, from Barr et al. 1985.)

Queens

Morphology

As dichthadiigynes, queens are blind, or nearly so; they possess a greatly enlarged gaster and single-segmented waist; they are permanently wingless; and they have strong legs (Figure 3.12). Correlated with this specialized reproductive design is an expanded tracheal system and the ability to store large amounts of fat as a reserve energy source (W. M. Wheeler 1928). The habitus of army ant queens uniquely adapts them to their specialized existence in the following ways:

1. The enlarged, ovary-filled gaster is well suited to the production of large numbers of eggs and thus to maintaining large colony size, which makes possible effective group predation.

2. Because the queen can produce large quantities of eggs in a short period, she is well adapted to a nomadic existence.

3. At times of emigration, the queen, by virtue of her well-developed legs, can move from one nesting site to another under her own power.

4. The development of the dichthadiigyne morphology, including the loss of wings, has compromised queens' ability to act as single foun-

dresses of new colonies and has led to fission as the source of new colonies (Gotwald 1987).

This list is not meant to suggest that the army ant queen habitus was arrived at through the sequential acquisition of these traits. No doubt this morphology was the result of reciprocally reinforcing, collateral changes that enhanced the queen's, as well as the colony's, survival.

Old World army ant queens lack eyes; New World species have a compound eye, but it is reduced to a single, ocellus-like, facet. The queen's mandible is linear and slightly curved apically and lacks subapical teeth. Among the Aenictini, Dorylini, and Ecitonini (the queen is unknown in the Cheliomyrmecini), the maxillary palpus has two segments, whereas the labial palpus is two-segmented in the ecitonines and one-segmented in the aenictines and dorylines (Gotwald 1969).

The mesosoma of the army ant queen is characterized by a general reduction in suturing, no doubt correlated with the loss of wings and flight in dichthadiigynes. Yet the queen's mesosoma is by no means uniform in all army ants. Queens of African *Aenictus*, for instance, have a simplified, or derived, thorax, whereas Asian species possess a complex, or more primitive, suturing (Gotwald and Cunningham–van Someren 1976).

The absence of wings and the simplified, workerlike mesosoma place the army ant queen among the category of reproductives called ergatoid (that is, workerlike) queens. Because her gaster is greatly enlarged, she is further regarded as a dichthadiiform ergatogyne (Hölldobler and Wilson 1990). Ergatoidy is a rather radical deviation from the morphology of queens that found new colonies alone, and it appears to be a major step in the evolution of army ant behavior (Villet 1989).

In the Ecitonini and Dorylini, the queen's petiole is enlarged and armed with caudally directed dorsal or lateral horns. These horns are most prominent in the ecitonines and apparently play a role in mating. In two *Eciton* matings described by Schneirla (1949), the males grasped the queen's petiolar horn during copulation. Such mandible-petiole contact may be consequential in mating and may serve as a species-isolating mechanism, especially given the fact that the petioles of queens and the mandibles of males differ so dramatically from species to species (Rettenmeyer 1963b). Additionally, Schneirla (1971) hypothesized that the petiolar horns serve a protective function.

The extreme expansiveness of the queen's gaster is a function of ovarian activity. During egg laying, when the ovaries hypertrophy, the gaster swells in a condition called physogastry. This occurs to a remarkable degree in phasic species of *Eciton*. During the statary phase, the inter-

Figure 3.13. Physogastric *Eciton burchelli* queen laying eggs. (Photograph by Carl W. Rettenmeyer, Connecticut Museum of Natural History.)

segmental membranes of the gaster stretch as they accommodate the enlarging ovaries, and the sclerites of the gaster separate from one another, forming islands in a sea of membrane (Figure 3.13) (Schneirla 1971).

During the nomadic phase, when the *Eciton* queen must perambulate from one nest site to another each and every night, the gaster is contracted; its five visible segments are strongly telescoped, and its sclerites overlap one another (Hagan 1954a). In *Eciton*, then, the physogastric condition is clearly synchronized with the alternation of the statary and nomadic phases of the colony. In nonphasic species, the cyclicity in physogastry is not so clearly documented. The gaster is rather uniquely developed in the Dorylinae, whose bifurcated hypopygium (the last ventral sclerite) extends prominently beyond the posterior margin of the pygidium (or last dorsal sclerite) (Figure 3.12) (Raignier and van Boven 1955, van Boven 1967).

Internal features of interest include a proventriculus that, in the *Eciton* queen, is smaller than in the worker, although identical in structure. The midgut of the *Eciton* queen, however, is different from the worker in both shape and histology. Additionally, the *Eciton* queen has approximately 30 Malpighian tubules, almost twice as many as the worker and no doubt a reflection of the queen's grandiose proportions. The elliptical or nearly round rectal pads are usually six in number, infrequently three (Whelden 1963). Except for the work of R. M. Whelden (1963) and H. R. Hagen (1954a–c) on *Eciton*, almost nothing is known of the internal anatomy of the army ant queen.

Whelden (1963) described numerous glands in the *Eciton* queen, in-

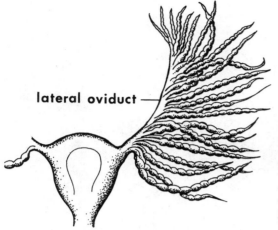

lateral oviduct

Figure 3.14. The ovary of a *Neivamyrmex* queen. The lateral oviduct extends the entire length of the ovary with individual polytrophic ovarioles opening into the oviduct along its length. (Redrawn from Holliday 1904.)

cluding mandibular, maxillary, and pharyngeal glands and three small, nameless glands that open through a membrane that extends from the mandible to the base of the mouthparts. The mesosoma contains labial glands, metapleural glands, a small gland at the base of each leg, and a moderately large, unnamed gland not present in workers (Whelden 1963). He detected, as well, a gland in the petiole and paired glands in each of the gaster's segments that open through the intersegmental membranes. The latter glands appear to be similar to the tergal glands of the workers but are impressively greater in number.

Supposedly, this massive array of exocrine glands in the gaster plays a significant role in the queen's chemical control of and substantial attractiveness to her workers (Watkins and Cole 1966). Franks and Hölldobler (1987, p. 236), preparing specimens for microscopic study, noted that secretions produced by these glands oozed from the glandular openings "like toothpaste squeezed out of a tube." Also present in the gaster are the poison sac and the Dufour's gland.

The reproductive system of *Eciton* and *Neivamyrmex* queens includes a vagina that opens into a median oviduct, which in turn bifurcates, forming lateral oviducts. These extend to paired ovaries composed of polytrophic ovarioles (Holliday 1904, Hagan 1954a, Whelden 1963). In *Neivamyrmex*, each lateral oviduct extends the entire length of its ovary, with the ovarioles opening into the oviduct along its length (Figure 3.14); in *Eciton*, each oviduct expands within its respective ovary to form a calyx capable of accommodating the vast numbers of oocytes that are discharged simultaneously into the oviduct (Figure 3.15). There are ap-

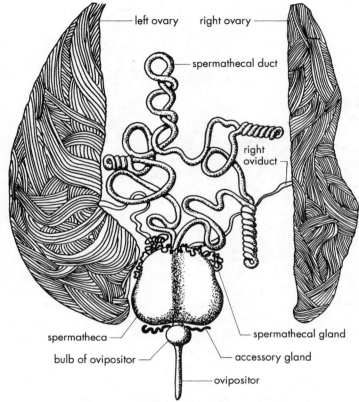

left ovary   right ovary

spermathecal duct

right
oviduct

spermatheca

bulb of ovipositor

spermathecal gland

accessory gland

ovipositor

Figure 3.15. Reproductive system of the *Eciton burchelli* queen (the ovaries' magnification is half that of the other parts). (Redrawn from Hagan 1954a, courtesy of the American Museum of Natural History Library.)

proximately 1200 ovarioles in the ovary of the *Eciton* queen (Hagan 1954a) and about 500 in the ovary of *Neivamyrmex* (Holliday 1904).

The *Eciton* queen has an exceptionally long spermathecal duct, which forms several tight coils that are peculiar in shape and position in each individual. This duct leads to a spherical or irregularly ovoid spermatheca, a structure common to all female insects that stores sperm from the time of insemination until the eggs are fertilized. Unevenly coiled, tubular spermathecal glands and a pair of accessory glands are also present in the *Eciton* queen (Hagan 1954a, Whelden 1963) (Figure 3.15).

Queen Functions
As the reproductive nucleus of the colony, the army ant queen plays a crucial role in maintaining the colony as an integrated social unit.

Without her, the colony would plummet irrevocably into disorder and extinction. Ordinarily, a colony that loses its queen cannot replace her unless a sexual brood is already present. Otherwise, the queenless colony's only hope of salvation lies in merging with a conspecific colony. A queenless colony of *Eciton*, for instance, will unite with a queen-normal colony within one or two hours of meeting it. The brood of this adopted colony, however, is consumed within a day or two of the merger (Schneirla 1971). Although the *Eciton* queen is a critical part of the functional pattern of the colony, she does not directly lead the business of the colony (Schneirla 1953). Rather, she is the "pacemaker" determining the colony's behavior patterns (Schneirla 1944).

With one known exception, army ants are exclusively monogynous; that is, each colony has but a single queen. This is hardly surprising. As Hölldobler and Wilson (1977) argued, the basic characteristics of insect society organization prejudice the evolution of species toward monogyny. This tendency, they proposed, is reversed only when special ecological limitations are placed on the species. Certainly, the presence of a single queen in army ants ensures the efficient regulation and coordination of worker behavior, a prerequisite to massive group foraging and emigration. Only in *Neivamyrmex carolinensis* do colonies regularly have multiple queens (Rettenmeyer and Watkins 1978), a condition that may relate to the harsh winters some colonies must endure. Under such marginal conditions, the probability of queen death may be considerable and ancillary queens may provide insurance against this frigid harbinger of colony extinction.

Perhaps the most unique functional feature of the queen is her role in colony founding. Unlike the great majority of ant queens, which found new colonies by themselves, the army ant queen acts in concert with her daughters to found new colonies by swarming, a process of colony fission also referred to as budding, hesmosis, and sociotomy (Hölldobler and Wilson 1977). Natural selection, however, should favor independent colony founding (Hölldobler and Wilson 1977), because mother colonies can create far more new colonies scattered over greater distances by producing foundress queens, each establishing a new colony, than they can by swarming. Hölldobler and Wilson concluded that swarming drains off a significant portion of the worker force from the original colony, and that the dispersal range of the new colony thus founded is limited by the challenges presented by group orientation and mobility. Swarming, they pointed out, is advantageous when the probability of the queen's survival is enhanced in the presence of workers, as measured against the survival of queens that go it alone. In the evolution of army ants, however, queen safety did not drive selection—the requirements of group predation did. Colony fission no doubt creates instant colonies of sufficient size to efficiently forage and retrieve prey en masse.

Like termite queens of the genus *Macrotermes,* the egg-laying capacity of some army ant queens can be truly extraordinary. A single queen of *Aenictus gracilis* may produce as many as 240,000 eggs per year; the queen of *Eciton burchelli* perhaps 2.4 million (Schneirla 1971). Impressive as these numbers are, they pale when compared with those for *Dorylus (Anomma) wilverthi,* whose queen, in the domain of oviposition, is unsurpassed. Raignier and van Boven (1955) estimated her egg production to be about 3 million to 4 million eggs per month, an annual yield of 36 to 48 million.

Although the queen influences the colony's behavior by the types of eggs she produces—that is, worker eggs or sexual brood eggs—the egg-laying schedule is determined by intracolony processes external to the queen (Schneirla 1971). Paramount in this "colony-situation-feedback" hypothesis is the effect of brood-stimulative phenomena that regulate the colony's activities through the mediation of the queen's corpora allata—paired glandular bodies, located behind the brain, that produce juvenile hormone. The corpora aliata play a major role in controlling normal oocyte development in most insects. Supposedly, according to Schneirla, the brood–corpora allata connection accounts for the queen's egg-laying cycles. This may be true, at least, for phasic army ant species that exhibit a well-marked functional cycle of regularly alternating nomadic and statary phases. Hagan (1954c, p. 19) found that the oocyte cycle in *Eciton* "harmonizes perfectly with colony behavior."

This so-called harmony is evident during the nomadic phase, when colonies emigrate to new nest sites on a nightly basis, in the species of *Eciton, Neivamyrmex,* and *Aenictus* studied by Schneirla (1971). At this time the queen's gaster remains contracted, a condition that unmistakably facilitates her transmigration over the rugged relief of the nomadic trail. As the nomadic phase ends and the statary, or quiescent, phase begins, the queen's gaster swells, exposing the intersegmental membranes. This distension continues for about a week into the statary phase, during which the colony remains in the same nest, until the queen achieves full physogastry and egg laying commences. The adaptive advantage of cyclic physogastry that coincides with the statary phase rests in minimizing risks to the queen during emigration. Surely, emigrating with her delicate intersegmental membranes exposed would subject the queen to undue peril.

The majority of army ant species do not have a well-defined functional cycle, however, and emigrations occur as single events, often separated by lengthy intervals. The queens of *Dorylus,* for example, do not follow a precise reproductive schedule (Raignier and van Boven 1955), and the period between emigrations may range from 3 to 45 days or more (Gotwald and Cunningham–van Someren 1990). Factors other than brood development apparently serve as proximate cues for emigration. The frequency and direction of emigration may be influenced by the amount and location of food, as they are in the ecitonine *Neivamyrmex nigrescens* (Topoff and Mirenda 1980a). In some species of *Dorylus*, the queen is in a permanent but moderate state of physogastry, although egg laying may be discontinuous. In *D. (A.) wilverthi*, egg-laying peaks at 20- to 25-day intervals and usually intensifies just after emigration (Raignier and van Boven 1955). A period of physogastry longer than that exhibited by *Eciton* queens may also exist in *Labidus*, a decidedly subterranean genus that apparently produces asynchronous broods (Rettenmeyer 1963b).

By insect standards, the army ant queen is undeniably a centenarian. It is not unusual for her to carry out her colony functions for two or more years. Indeed, one *E. hamatum* queen is known to have functioned for at least four and a half years. During this time she may have produced about 50 all-worker broods (or about 4 million workers) and one or more sexual broods (Schneirla 1971). Schneirla (1971) speculated that most *Eciton* queens meet their demise when superseded by virgin queens at the time of colony division. For the most part, however, the longevity of army ant queens has been ignored. What is known about queen survival is based on the recovery of marked queens in the field or on laboratory colony observations. Tschinkel (1987) was able to estimate the age of a fire ant (*Solenopsis invicta*) queen by examining the sperm content of her spermatheca (it diminishes with age), but this method is practical only for queens that mate but once in their lives. For army ant queens, which apparently engage in multiple matings, another method will have to be devised.

Males

Morphology

As noted in Chapter 1, the morphology of the army ant male is unique, with its capacious gaster, modified mandibles, and uncommonly well-developed genitalia (Figure 3.16). Strangely, the male approximates in size and morphology its conspecific queen, a glaring departure from most ant males, which are puny, at best, compared with their queens. Franks and Hölldobler (1987) suggested that convergence in external morphology—especially in exocrine glands in males and queens—begs the conclusion that males mimic queens in order to be accepted by the workers of the conspecific colonies they must enter if they are to mate.

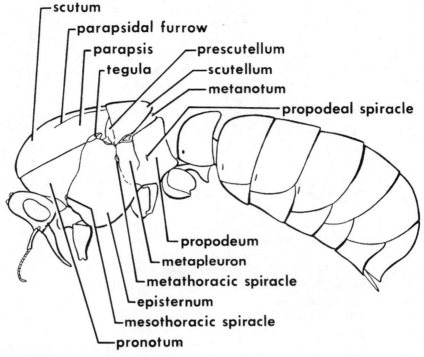

Figure 3.16. External features of the *Dorylus* (*Anomma*) male, legs and wings omitted. (Reprinted, by permission of the National Research Council of Canada, from Barr and Gotwald 1982, *Canadian Journal of Zoology* 60:2652–2658.)

They concluded, however, that sexual selection, acting on the competitive success of males and queens and on the ability of workers to discriminate, produced a convergence in the way both males and females demonstrate their potential fertility to workers. In other words, the workers select their queen from a batch of new queens at time of colony fission and then decide which alien males they will admit to their nest to inseminate her (see Chapter 4). The workers must glean the best in this fertility contest; the males and the queen must demonstrate themselves worthy of selection. Perhaps the similarity between males and females is not so much the result of convergence as it is the product of homology. That is, males and females might share structures of identical embryonic origin that are unrelated in function.

The male's conspicuous compound eyes and ocelli no doubt serve it well when it emerges from its mother colony to take flight in search of a receptive conspecific colony. The mandibles are distinctive enough to

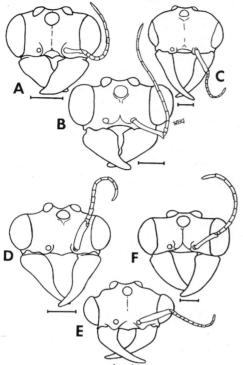

Figure 3.17. Heads of *Dorylus* males, showing the formidable mandibles, which are distinctive enough to be of value in separating the subgenera. Subgenera *Alaopone* (A), *Dichthadia* (B), *Anomma* (C), *Dorylus* (D), *Typhlopone* (E), and *Rhogmus* (F). Each scale equals 1 mm. (Reprinted, by permission of the National Research Council of Canada, from Barr and Gotwald 1982, *Canadian Journal of Zoology* 60:2652–2658.)

be of practical value in separating the subgenera of *Dorylus* (Figure 3.17) (Gotwald 1969). Observations indicate that the mandibles are used to grasp the female during copulation (Schneirla 1949, Rettenmeyer 1963b). Certainly, the impressive mandibles pose no defensive threat to potential predators, as the males are weak biters. I have frequently collected males with impunity and have never been bitten. The maxillary palpus of the male has two segments in all four tribes, but the labial palpus is one-segmented in the Aenictini and Dorylini, two-segmented in the Ecitonini, and three-segmented in the Cheliomyrmecini (Gotwald 1969).

G. S. Tulloch (1935) regarded the male mesosoma as highly specialized and noted that the alitrunk of *Dorylus* (*D.*) *helvolus* represented the greatest departure from the fundamental type found in other subfamilies. Venation in the wings of male army ants is not reduced and is therefore primitive (W. L. Brown and Nutting 1950). The wings are most primitively veined in *Cheliomyrmex*, but little else can be said, since wing venation patterns in both subfamilies await renewed analysis. The male's

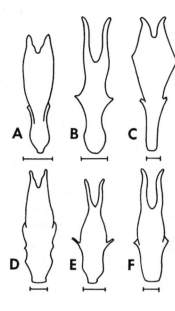

Figure 3.18. Subgenital plates of *Dorylus* males. The forked nature of this plate, the sternite of the ninth abdominal segment, constitutes a synapomorphic character of the doryline section of ant subfamilies. Subgenera *Alaopone* (A), *Dichthadia* (B), *Rhogmus* (C), *Typhlopone* (D), *Dorylus* (E), and *Anomma* (F). Each scale equals 1 mm. (Redrawn, by permission of the National Research Council of Canada, from Barr and Gotwald 1982, *Canadian Journal of Zoology* 60:2652–2658.)

waist is one-segmented, as is the queen's, but unlike the queen's, it is unarmed.

The sturdy genitalic capsule of the male is retracted into a large cavity ventral to the rectum and anus in the last few gastral segments. Of the sclerites composing the genitalia, only one, the forked subgenital plate (sternite of the ninth abdominal segment), projects beyond the tip of the gaster (Figure 3.18). Bolton (1990c, p. 1351) recognized that this "biaculeate" plate constitutes a synapomorphic character of the doryline section of ant subfamilies. Curiously, when captured, the male maneuvers and orients its gaster as if it were stinging, a disconcerting behavior to even the boldest of human captors. As in other ants, the external genitalia consist of three pairs of valves surrounded anteriorly by a basal ring (Forbes and Do-Van-Quy 1965). The outer valves are generally called the parameres, the middle valves the volsellae, and the inner valves the aedeagus. The basal ring is manifestly wider in the ecitonines than in the dorylines (Forbes 1958).

The alimentary canal of the male, at least in *Aenictus gracilis*, includes a pharynx, esophagus, cardiac valve, ventriculus, pylorus, intestine, rectal valve, and rectum. Notably absent in this species and in *Dorylus* (*Typhlopone*) *labiatus*, the only other species whose digestive tract has been examined, is the crop and its associated proventriculus (Mukerjee 1926, Shyamalanath and Forbes 1980). The absence of the crop should

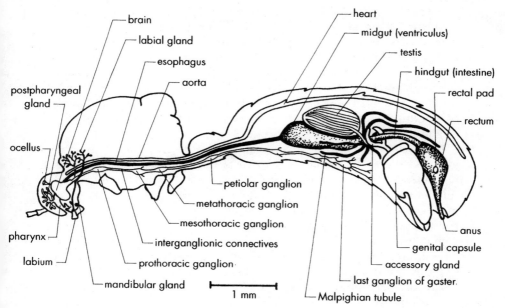

Figure 3.19. Internal morphology of the *Aenictus gracilis* male. (Redrawn, by permission of James Forbes and the New York Entomological Society, from Shyamalanath and Forbes 1980.)

be anticipated since males do not forage or gather food. The *A. gracilis* male commonly possesses 18 to 20 Malpighian tubules and three rectal pads, two located laterally and one ventrally (Figure 3.19).

The glandular anatomy of *A. gracilis,* and by implication other army ant males, includes, in the head, the postpharyngeal glands and a pair of mandibular glands located at the bases of the mandibles. Maxillary glands are absent. Labial (salivary) glands are present in the anterior mesosoma but are noticeably small (Shyamalanath and Forbes 1980). The mesosoma in all male army ants is devoid of the metapleural gland. W. L. Brown (1968) speculated that metapleural glands produce a substance that labels all the members of a colony and marks as aliens all outsiders that attempt to enter. He further postulated that, since the army ant male must mate with a queen in a colony other than its own, males lacking this labeling substance can enter foreign conspecific colonies unmolested.

In the gaster of *Neivamyrmex* and *Eciton* there are two major glandular structures associated with the genitalia. The first of these, the aedeagal, or penis, glands, are located inside the aedeagus. The second are paired clusters of glandular cells located in the ninth sternite that are called the subgenital plate gland (Hölldobler and Engel-Siegel 1982). These two genera also possess large pygidial glands that open between tergites 6 and 7 and similar tergal glands that broach the intersegmental membranes between tergites 6 and 5, 5 and 4, and 4 and 3. Additionally, intersegmental glandular cells are found between the seventh and eighth tergites. And most unusually, in the third tergite there are paired groups of glandular cells whose ducts pierce the sclerotized cuticle of the third tergite (Hölldobler and Engel-Siegel 1982).

Ecitonine males also have an abundance of sternal glands, a curious gland in the third sternite whose ducts pass through the cuticle, and anus glands (Hölldobler and Engel-Siegel 1982). The males are markedly different from the workers in being richly endowed with abdominal glands, much like those found in queens. It was this similarity that gave rise to the hypothesis that males imitate queens chemically and thereby gain entrance to foreign conspecific colonies and their queens (Hölldobler and Engel-Siegel 1982). This idea of the ambrosial suitor is quite the opposite of Brown's sanitized, odorless male.

The internal reproductive system of male ants includes a pair of testes, each composed of a variable number of follicles in which spermatogenesis occurs; the vasa deferentia, narrow tubes that transport the spermatozoa from the testes to the middle or base of the accessory glands, the ejaculatory duct, or both; and the accessory glands, which probably contribute to the formation of seminal fluid. The vasa deferentia are, in some species, swollen or bulbous and store sperm cells; in this case they are referred to as seminal vesicles. Ducts leading from the accessory glands fuse to form the ejaculatory duct, although within the duct their lumina remain separate and distinct (Gotwald and Burdette 1981).

The internal reproductive system of male army ants is noteworthy in two respects: (1) functional testes, for the most part, are present only in the pupa—that is, spermatogenesis occurs during pupal development, with the testes rapidly degenerating after the male emerges as an adult; and (2) the vasa deferentia are always enlarged to form seminal vesicles. Although Gotwald and Burdette (1981) failed to find testes in adult males of *Cheliomyrmex*, *Labidus*, *Eciton*, and *Dorylus* and observed only atrophied testicular follicles in *Neivamyrmex* and *Nomamyrmex*, other researchers have discovered prominent testes in adult *E. hamatum* (Forbes 1958), *A. gracilis* (Shyamalanath and Forbes 1983), and *D. (A.) wilverthi* (Ford and Forbes 1980, 1983). This discrepency may reflect the ages of the males dissected, since the state of testicular degeneration is probably a function of time.

The number of follicles composing each testis ranges from 20 in *E.*

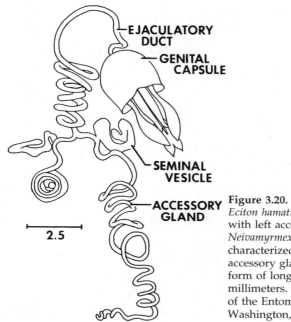

EJACULATORY DUCT

GENITAL CAPSULE

SEMINAL VESICLE

ACCESSORY GLAND

2.5

Figure 3.20. Reproductive system of *Eciton hamatum* male, dorsal view, with left accessory gland uncoiled. *Neivamyrmex* and *Eciton* males are characterized by their extraordinary accessory glands, which take the form of long, coiled tubes. Scale in millimeters. (Reprinted, by courtesy of the Entomological Society of Washington, from Gotwald and Burdette 1981.)

hamatum and 22–25 in *Neivamyrmex harrisi* to 35–40 in *D. (A.) wilverthi* and 50–55 in *D. (A.) nigricans* (Ford and Forbes 1980). The seminal vesicles are saclike in all genera, although they assume different configurations (Figures 3.20, 3.21). The accessory glands, on the other hand, are extraordinary in *Neivamyrmex* and *Eciton,* in which they take the form of long, coiled tubes, a development apparently unique among the ants (Hung and Vinson 1975). Gotwald and Burdette (1981) concluded that the internal reproductive system is most derived in the tribe Ecitonini and least derived in the Dorylinae; among the New World army ants, *Cheliomyrmex* is the most primitive.

Male Functions

The mission of the army ant male appears straightforward enough: take flight from the parent colony, locate a conspecific colony, win the acceptance of the foreign workers and thereby entrance to their nest, and inseminate the colony's queen. But the mission is fraught with logistical challenges and a multitude of hazards.

Newly enclosed adult males contribute nothing to the daily maintenance of their parent colony. Their role as utilitarian, mobile sperm do-

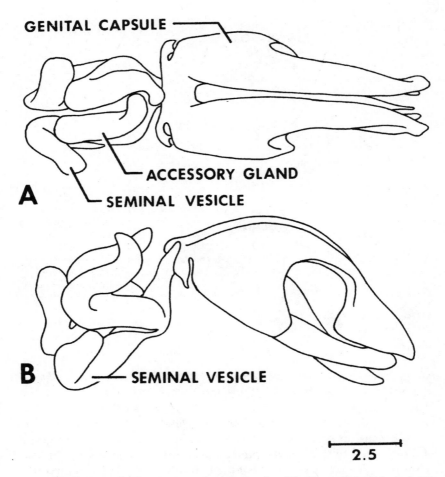

Figure 3.21. Reproductive system of male *Dorylus* (*Typhlopone*), dorsal view (A) and lateral view (B). Scale in millimeters. (Reprinted, by courtesy of the Entomological Society of Washington, from Gotwald and Burdette 1981.)

nors is important, to be sure, but they have little else to look forward to but death. Although this sounds like so much anthropomorphic melodrama, I believe it portrays the male's life vividly and accurately. If death does not come in the clutches of a ravenous predator (Plates 18A and 19), it will most certainly arrive soon after the alien queen is inseminated, for the successful male is a has-been.

In *Eciton* and other genera whose colonies pass through alternating statary and nomadic phases, males are produced periodically in large

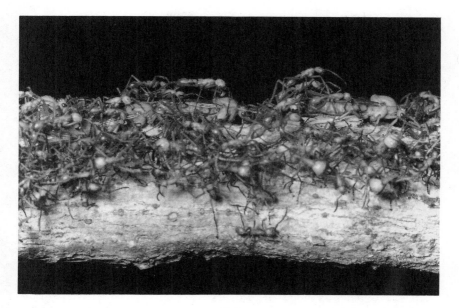

Figure 3.22. Males in an emigration column of *Eciton hamatum*. (Photograph by
Carl W. Rettenmeyer, Connecticut Museum of Natural History.)

sexual broods (which include queens), which herald imminent colony
division. The sexual broods are precisely coordinated with the all-worker
broods that precede and follow them (Schneirla 1971). This precision in
the appearance of males is probably absent in nonphasic army ant species. In the driver ants of the genus *Dorylus,* for instance, sexual broods
can appear during any season (Raignier 1959, 1972). Therefore, males'
exodus from the nest following their eclosion is relatively precise and
genus-typical in species with a regular functional cycle (Schneirla 1971)
and variable in other groups (Gotwald 1982).

   Although males may fly from their parent nest soon after eclosion,
they sometimes, in some species, emigrate with their colony. In *Eciton,*
this occurs during a new daughter colony's first nomadic phase following colony division (Figure 3.22). The males may separate from the
workers and fly away as the column proceeds along the emigration
route, and in this way they "literally seed the area through which they
pass" (Schneirla 1971, p. 253). In the driver ants, fully grown male larvae
and pupae are left behind in the old nest after colony division; the pupae
then eclose and the adult males take flight (Raignier 1972).

   A successful exodus flight ends with the discovery of an alien conspecific colony. Raignier and van Boven (1955) reported that *Anomma* driver ant males locate the abandoned chemical trails of other colonies and follow them to the nest site, and Schneirla (1971) conjectured that postflight *Eciton* males may attract conspecific workers by releasing a pheromone attractant that they spread on the substrate with a brushlike collection of setae at the tip of the gaster. Following the exodus flight the males lose their wings, possibly as a physiological consequence of the flight itself, although they may be torn off by the workers of the adoptive colony.

The presence of wingless males moving in columns of workers was initially confusing to army ant observers. In 1849, for instance, Dr. Thomas S. Savage (1849, p. 201), on missionary assignment in "Gaboon," in west-central Africa, puzzled over the wingless males he collected from a column of driver ants but concluded correctly that "they seemed to be no unimportant members of the community." In 1938, V. G. L. van Someren of Nairobi made similar observations of dealate males, unaware of the recorded accounts of Savage. These sightings of dealate males were discussed and reviewed by H. S. J. K. Donisthorpe (1939), who viewed the males as enigmatic wooers.

Of equal interest are the diurnal/nocturnal flight periodicities of the army ant males. Schneirla (1971) suggested that males of surface-active species leave their colonies around dusk and those of subterranean species exit in the evening or at night. Some surface-active species, however, such as *D. (A.) nigricans* and *D. (A.) wilverthi*, fly in complete darkness, usually after 10:00 P.M. (J. K. A. van Boven, pers. comm.).

Schneirla additionally noted that males of different species may respond differently to environmental stimuli once they have landed following their exodus flight. For example, surface-active species of *Neivamyrmex*, which have relatively small, flat eyes, react positively to ecitonine chemical trails and may locate conspecific colonies through chemical stimuli. On the other hand, subterranean species have large eyes, react weakly to chemical trails, and may rely on visual stimuli, moving toward moonlight-silhouetted objects such as rocks and logs where the trails of subterranean species are most likely to be located (Schneirla 1971).

Indeed, a correlation between eye size and circadian flight periodicity has been confirmed, at least in the genus *Neivamyrmex*, but the relationship between eye size and subterranean and surface-active lifeways awaits empirical verification (Coody and Watkins 1986). The ocelli and compound eyes of *Neivamyrmex* males of three night-flying species were significantly larger than those of three day-flying species examined by C. J. Coody and J. F. Watkins (1986).

Rettenmeyer (1963b) concluded that New World males are negatively phototaxic and unable to fly for one to three days after eclosion. He proposed that only after flight do the males become positively phototaxic—that is, attracted to light. P. B. Kannowski (1969) examined the frequency distribution of ecitonine males during the nighttime hours and discovered that the species form two distinct groups: one that conducts postsunset flights and another that launches predawn flights. One species was "circum-nocturnal." A. J. Haddow et al. (1966) found that the males of eight species of *Dorylus* (representing five subgenera) from Uganda peak in abundance at different times during the night (Figure 3.23). That is, each species has a well-defined flight time. The adaptive significance of this succession of flight times is unclear. A similar study of four *Neivamyrmex* species in central Texas did not find clearly defined species-specific flight peaks (Baldridge et al. 1980). The males of several species of *Neivamyrmex* fly during the day (Baldridge et al. 1980).

Some army ant species also exhibit seasonal flight periodicity. Although Haddow and his associates (1966) concluded that *Dorylus* males fly at all times of the year, Dennis Leston (1979), well known for his elucidation of the cocoa canopy ant mosaic, discovered a regular cycle in the timing of *Dorylus* male flights. He noted, for instance, that males are produced in numerous colonies in a given locale about every 30 to 32 days from March through September and around every 28 or 29 days from December through February (Figure 3.24). This cycle is synchronous in the four most commonly trapped species. Males are produced more or less throughout the year but with distinct seasonal peaks (Leston 1979). Leston concluded that the syncronicity evident in these cycles is not related to climatic factors. Instead, he speculated, it is a mechanism for oversaturating an area with males, so that at least some males would survive an onslaught of predators. The flights of New World army ant males are clearly more seasonal than those of *Dorylus* (M. R. Smith 1942, Borgmeier 1955, Rettenmeyer 1963b, Kannowski 1969, Baldridge et al. 1980, Watkins 1985). In general, these males fly during the dry season and the early rainy season.

Nothing is known of males' success in inseminating the queens of foreign colonies or if males ever—frequently or infrequently—copulate with their mother queen or newly emerged sister queens. Once in his adoptive colony, a dealate male may remain for days or even weeks before he mates and dies (Schneirla 1971). Rettenmeyer (1963b) supposed that such males probably live a few days but seldom more than three weeks.

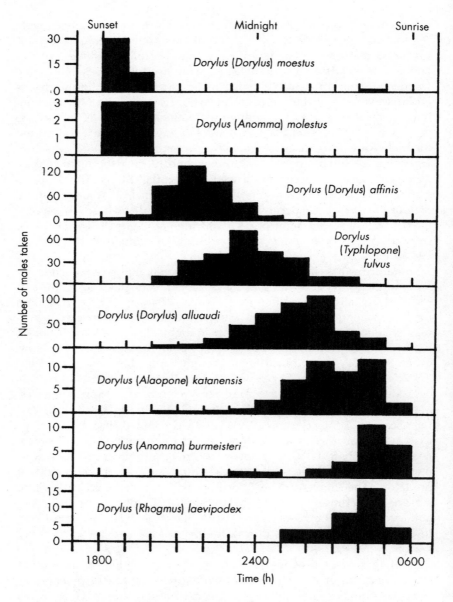

Figure 3.23. Daily flight periodicities of *Dorylus* males collected in light traps in East Africa. The data suggest that different species take flight at different times, resulting in the temporal spacing of species. (Redrawn, by permission of the Royal Entomological Society, from Haddow et al. 1966.)

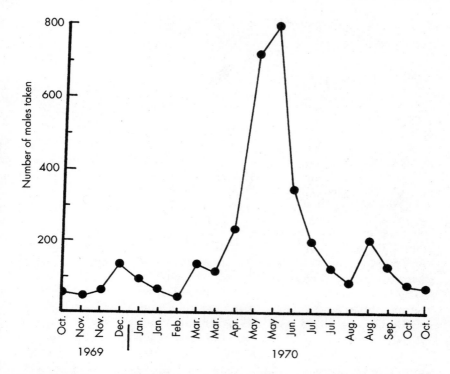

Figure 3.24. Seasonal flight periodicities of *Dorylus* males collected in light traps in Ghana. Although males of different species are produced more or less throughout the year, they have synchronized seasonal peaks which may serve as a mechanism to oversaturate the area and thus enhance their chance of surviving predators. (Redrawn, by permission of the Cambridge Entomological Club, from Leston 1979.)

Brood

Morphology

Of the brood, pupae, deceptively inert in the army ant scheme of things, have attracted the least attention. Larvae, on the other hand, possessing intrinsic interest for the taxonomist, have garnered more than their share of probing investigations (see G. C. Wheeler 1943; Tafuri 1955; Lappano 1958; G. C. Wheeler and Wheeler 1964, 1974, 1984, 1986; Schneirla et al. 1968). Consequently, given the dearth of information on postlarval development, pupae will be all but ignored here.

Army ant worker larvae are elongate, rather slender, and curved ventrally (G. C. Wheeler and Wheeler 1984). So-called vestigial legs are con-

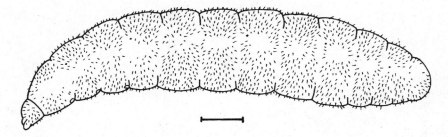

Figure 3.25. *Eciton hamatum* worker larva. Army ant worker larvae have an elongate, rather slender body of 12 or 13 distinct segments. Scale equals 1 mm. (Redrawn, by permission of J. N. Wheeler and the Kansas Entomological Society, from G. C. Wheeler and Wheeler 1984.)

spicuous, and each is associated with an imaginal disc, a patch of embryonic tissue typical of insects with complete metamorphosis that gives rise to adult structures. The larval body is composed of 12 or 13 distinct segments (Figure 3.25) (G. C. Wheeler 1938).

The larvae possess relatively large antennae, each with two sensilla. The mouthparts are small, and the mandibles are essentially vestigial and lightly sclerotized (G. C. Wheeler and Wheeler 1984). George C. Wheeler and Jeanette Wheeler, who devoted a lifetime to the study of ant larvae, find this paradoxical. "Why," they asked, "do larvae of the world's most carnivorous ants have mandibles incapable of chewing solid food?" (Wheeler and Wheeler 1984, p. 274).

W. M. Wheeler and I. W. Bailey (1925) suspected that *Eciton burchelli* larvae are fed large pellets composed of the rolled-up soft parts of prey. The pellets they examined were so compact that they retained their shape even in the larval gut, where they formed an irregular elongate series. Furthermore, Wheeler and Bailey concluded, the workers preparing these pellets must trim away the hard chitinous materials, and may as well consume the liquids expressed from the prey tissue as it is compressed. R. S. Petralia and S. B. Vinson (1979) noted that the larvae of *Neivamyrmex nigrescens* do not possess specialized structures for holding food on the ventral body region as do other ant species. Apparently, then, army ant larvae are not required to hold or manipulate their food. Schneirla (1971, p. 140) observed larval feeding in *Eciton*:

> Morsels of booty, macerated as they are rolled about and squeezed between the jaws, tongue, and palps of the adult workers, become soft pellets which are laid upon larvae or larvae are laid upon them. The process of food preparation often begins with workers pulling against each other in pairs with a piece of booty held between them in their jaws,

evidently extracting juices with each new grip and pull. In keeping with the workers' degenerate crops and small gasters, their own food may be the fluids more often than the tissues. The larvae, in contrast, feed voraciously and are often seen with their mouthparts applied to morsels or—increasingly as they grow—to whole pieces of booty.

The Wheelers (1984) conjectured that the larvae excrete, from the mouth, an enzyme that digests the solid prey tissues and that the resultant fluid is then drawn into the pharynx. Larval mouthpart morphology, they noted, is consistent with the ability to aspirate liquids.

As is the case with all immature insects, the larvae grow through a series of successive molts. That is, the exoskeleton limits growth and must be replaced periodically by a new, expandable cuticle, following which the old cuticle is shed, like an undersized shirt popping its buttons. Each stage between molts is an instar, and the number of instars in ants is generally species specific. *E. burchelli* and *E. hamatum* appear to have five instars, the fifth representing the mature larva (G. C. Wheeler and Wheeler 1986). The number of instars found in other ecitonines and in the aenictines and dorylines is not known.

Queen and male larvae are present in the colony only periodically and have not been the subjects of detailed morphological investigations, perhaps because of their scarcity in collections (Figure 3.26). The mature male larva of *D. (A.) molestus*, the East African driver ant, can only be described as colossal in comparison with the worker larva (G. C. Wheeler 1943). Most distinctively, the anterior portion of this larva is bent ventrally at an angle of 90°. Raignier (1972) recorded that the young male driver ant larva one week old or older can be distinguished from the worker larva because it is slender and makes curling movements. According to Raignier, these movements may provoke increased foraging activity among the adult workers.

The internal anatomy of army ant larvae has been detailed for two species, *E. burchelli* (Lappano 1958) and *N. nigrescens* (Wang and Happ 1974). The alimentary canal of both species is essentially a straight tube composed of a foregut, midgut, and hindgut. The midgut of *E. burchelli* extends from the second thoracic segment to the eleventh segment as a large, dilated, blind sac (Lappano 1958). It is the largest organ in the larva. The hindgut begins where the midgut terminates and consists of a short, narrow intestine, an expanded rectum, and a constricted anus. Four Malpighian tubules insert at the anterior end of the intestine. The cavities of the midgut and hindgut become continuous in late larval or early pupal development. W. M. Wheeler and Bailey (1925) described the gut of *E. burchelli* larvae as unlike that of other known ant larvae

84   Army Ants

because it is quite long and slender, a description at variance with that of Lappano.

The nervous system of *E. burchelli* larvae is composed of a central division, consisting of a brain and central nerve cord with 12 paired ganglia, and a stomatogastric division. This species possesses a secretory system that includes corpora allata and labial glands. The labial glands, prominent as well in *N. nigrescens*, undergo striking morphological and histological changes during the course of the larva's development. They may play a major role in influencing colony activities through the regulatory substances they release. Certainly, increased physiological activity of the glands is correlated with heightened activity of the workers during the nomadic phase (Lappano 1958, Wang and Happ 1974). Undeveloped ovaries are discernible in *E. burchelli* larvae as paired, elon-

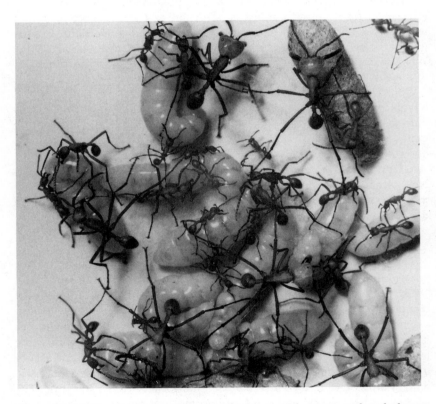

Figure 3.26. *Eciton hamatum* workers and soldiers with queen and male larvae. Sexual larvae are present in the army ant colony only periodically. (Photograph by Carl W. Rettenmeyer, Connecticut Museum of Natural History.)

gated bodies situated in the middorsal region of the ninth or tenth segments (Lappano 1958).

At the end of larval development, some species spin cocoons; others do not. A phyletic pattern in cocoon spinning is not apparent, although it is assumed that the ant cocoon was inherited from a wasplike ancestor (W. M. Wheeler 1915) and that the absence of a cocoon is therefore a derived characteristic. *Dorylus* and *Aenictus* larvae do not spin cocoons; those of *Eciton* and *Labidus* do (see Figure 3.26). Curiously, only the queen and male larvae of *Neivamyrmex* spin cocoons; the workers pupate naked. William Beebe (1919, p. 463) observed cocoon spinning in *Eciton* and wrote, "I watched the very first thread of silk drawn between the larva and the outside world, and in an incredibly short time the cocoon was outlined in a tissue-thin, transparent aura within which the tenant could be seen skillfully weaving its own shroud."

Brood Biology

Army ant colonies always contain brood, although in phasic species the brood are produced according to a schedule. Larvae are present in the nomadic phase, when the colony is energized to emigrate on a nightly basis, and pupae are present in the statary phase, when colony activity, including foraging, is low. In classic ethological studies, Schneirla (1933, 1938) determined that it is the larvae that stimulate the colony to its nomadic phase activity levels, and the pupae (or perhaps the absence of larvae) that induce the depressed activity levels of the statary phase.

His analysis of army ant behavior led Schneirla (1938) to his brood-stimulative theory of army ant cyclic activity. Initially working with *Eciton*, he demonstrated that the nomadic phase begins with the emergence of callow workers from their cocoons, which has a vitalizing effect on the colony. The nightly emigrations and frenzied daytime foraging of the nomadic phase end when the larvae mature and pupate. With pupation and the subsequent lack of stimulation from the brood, daily emigrations cease and the queen achieves full physogastry; the statary phase has begun. At the middle of this immotive phase, the queen deposits a single series of eggs that constitute a unitary population. These eggs hatch just before the emergence of the callow workers, which in combination initiate the next nomadic phase. These well-marked functional cycles have been identified in at least some species of *Eciton, Neivamyrmex,* and Asian *Aenictus* (Schneirla 1971).

The phasic species must be depicted as anomalous, however, for the great preponderance of army ant species are nonphasic. In *Dorylus,* for instance, and no doubt in numerous species in other genera, the functional cycles are irregular. Although Raignier and van Boven (1955) found that emigrations in *D. (A.) wilverthi* can be initiated by the eclosion of callow workers, 6- to 40-day intervals separate emigrations. One colony of the East African driver ant *D. (A.) molestus* that was followed for 432 consecutive days (the longest continuous observation of any Old World army ant colony; see Gotwald and Cunningham–van Someren 1990) emigrated 38 times, with the shortest stay at any nest site being 3 days, and the longest, 45 days. Factors other than brood development must serve as proximate cues for emigration in this and other nonphasic species.

Nest Structure

Although Schneirla (1971) applied the term *bivouac* to all army ant nests, implying through the use of this military metaphor that these nests are temporary, more "the state of the colony" than a physical place, the name is most certainly a misnomer when applied to subterranean nests. Indeed, all army ants are nomadic, but most species also reside in subterranean quarters for unpredictable lengths of time. Raignier and van Boven (1955), for example, recorded one intermigratory interval for *D. (A.) nigricans* of 125 days. The term *bivouac* is most appropriate for the surface or above-surface nests formed by phasic species, especially those of the genus *Eciton*. Constructed from the bodies of the ants themselves, these bivouacs hardly qualify as nests at all in the conventional sense. This duality in army ant nest formation was a challenge even to the great myrmecologist Auguste Forel (1896, p. 501), who, in his classification of ant nests, noted that some army ants have "earth nests" and others "migratory nests." He was unsure how to reconcile the two within a group that he considered homogeneous and monophyletic.

Raignier and van Boven (1955) found the subterranean nests of driver ants to be of two general types, one exemplified by *D. (A.) wilverthi* and the other by *D. (A.) nigricans*. In the first species, the entire colony clusters together in a central cavity, forming a compact mass in which a higher, presumably optimal, temperature can be maintained. This nest is 1 to 2 meters deep. The nest of *D. (A.) nigricans* has no central chamber. Colony members are dispersed in deep galleries and chambers, 2 to 4 meters down in the soil, and apparently have little ability to control nest temperature (Figure 3.27). Both nest types are usually constructed among the root systems of living trees. Their presence is indicated by characteristic craters composed of excavated soil particles on the sub-

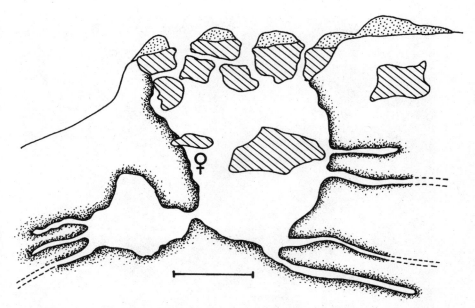

Figure 3.27. A *Dorylus* (*Anomma*) *nigricans* nest, consisting of a large central chamber, occupied by workers and larvae, that gives rise to numerous galleries leading to small satellite chambers. Craters of excavated soil mark the nest's location; the nest is constructed among small stones (crosshatching). Note the location of the queen. Scale equals 30 cm. (Redrawn, by permission of J. K. A. van Boven and the Université Nationale, Service des Publications, République de Côte d'Ivoire, from van Boven and Levieux 1968, *Annales de l'Université d'Abidjan* (Côte d'Ivoire), ser. E, 1(2):354.)

strate surface (Figure 3.28). In a study of *D.* (*A.*) *nigricans* in savanna and gallery forest in Ivory Coast, J. M. Leroux (1977a) concluded that during the first week of nest construction as much as 20 kg of soil a day may be excavated by the workers. The *nigricans* queen typically lodges deep inside the nest, whereas the *wilverthi* queen is variably located.

Driver ant nests are not highly organized dwellings. Often the ants use existing cavities and chambers in the soil, especially those associated with tree root systems. Savage (1849, p. 196) observed over a century ago that the interior of the driver ant nest "exhibits no mechanical contrivance for the depositing of food, or hatching of eggs; for these purposes, spaces between the stone, sticks, etc., found within, are adopted." Driver ants are not averse to using nests that they or other conspecific colonies have previously occupied. In a year-long study of a *D.* (*A.*) *molestus* colony, Gotwald and Cunningham–van Someren (1990) found

Figure 3.28. Soil particle surface craters of a *Dorylus* (*Anomma*) *wilverthi* nest. In the first week of nest construction, as much as 20 kg of soil a day may be excavated by driver ant workers. (Reprinted, by permission of A. Raignier and the Africa Museum, Tervuren, from Raignier and van Boven 1955.)

that the colony inhabited one nest site on ten separate occasions for a total of 170 days. The same nest site had also been utilized by three other colonies.

Dorylus species are strictly hypogaeic nesters; *Aenictus* species are not, although most do nest in the soil. Some New World army ants nest in the soil at some times and form bivouacs at other times. For example, *Eciton vagans* colonies may form more or less exposed bivouacs, but more often than not they lodge in underground or well-sheltered nests (Rettenmeyer 1963b). *Labidus coecus* nests may be located deep in the soil or close to the surface. Rettenmeyer (1963b) discovered one nest of this species in a decaying log. The eggs were concentrated in three long chambers, the cocoons were situated primarily in a separate cavity, and the larvae were scattered throughout the nest in numerous compartments. The nests of *Neivamyrmex nigrescens* are also subterranean, but their inhabitants conduct a daily vertical migration in which a portion of the colony's brood is brought near the surface to lie beneath sun-warmed stones (Rettenmeyer 1963b).

The true bivouac is a wondrous sight to behold. Although constructed wholly of the ants themselves, its structural integrity gives credence to the theory of the superorganism. It is like a single animal lying sequestered between the buttresses of a tree or suspended, slothlike, from some half-fallen tree trunk. Upon first encountering an *Eciton* bivouac, Beebe (1921, p. 60) marveled that "this chocolate-colored mass with its myriad ivory dots [the heads of major workers] was the home, the nest, the hearth, the nursery, the bridal suite, the kitchen, the bed and board of the army ants." He (p. 60) noted that Louis XIV, king of France and creator of Versailles, once exclaimed, "L'Etat, c'est moi!" and opined that "this figure of speech becomes an empty meaningless phrase beside what an army ant could boast—'La maison, c'est moi!' " Captivated by what he saw, Beebe (1921, pp. 60–61) wrote of the bivouac:

> Every rafter, beam, stringer, window-frame and door-frame, hall-way, room, ceiling, wall and floor, foundation, superstructure and roof, all were ants—living ants, distorted by stress, crowded into the dense walls, spread out to widest stretch across tie-spaces. I had thought it marvelous when I saw them arrange themselves as bridges, walks, hand-rails, buttresses, and sign-boards along the columns; but this new absorption of environment, this usurpation of wood and stone, this insinuation of themselves into the province of the inorganic world, was almost too astounding to credit.

The configuration and location of the bivouac depends on the phase in which the colony is engaged. When the colony is statary, the nesting cluster assembles in an enclosed or sheltered space where it remains for many days; when nomadic, a new exposed cluster is formed each night in a new location. *E. hamatum* bivouacs are seldom more than 1 meter above the ground, whereas the bivouac of *E. burchelli* may be formed as high as 30 meters in a tree (Schneirla 1971, Teles da Silva 1977). A bivouac is formed of the bodies of the ants themselves, suspended from a support object and from each other. Bivouac formation is made possible by a clustering response found even in the driver ants—which also, under certain circumstances, produce hanging clusters. In nomadic *E. hamatum*, bivouac formation begins at dusk when workers form clusters suspended from a log or other support object near a cache of prey. Additional workers are attracted to the clusters and attach themselves, usually by interlocking their tarsal claws. As W. M. Wheeler (1900, p. 569) fancifully wrote, in "forming these chains, which remind one of the pictures of prehensile-tailed monkeys crossing a stream, the insects make good use of their long legs and hooked claws." First strands, then

Figure 3.29. Surface of an *Eciton hamatum* bivouac formed of worker ants, attached to one another by interlocking tarsal claws, hanging heads downward. (Photograph by Carl W. Rettenmeyer, Connecticut Museum of Natural History.)

"ropes" of workers are formed, and these ultimately fuse into a "heavy fabric" (Schneirla 1971, p. 55). Workers integrated into this organic tapestry usually hang head downward, a phenomenon that Wheeler (1900, p. 569) attributed to the "positively geotropic" nature of the workers, but which Schneirla (1971) explained in terms of physical stresses exerted within the suspended curtain (Figure 3.29). The *E. hamatum* bivouac may take the form of a cylinder hanging between a support object and the substrate (Figure 3.30), or it may be a curtain of ants suspended between the buttresses of a tree (Plate 4) (Schneirla 1971). In *Aenictus gracilis* and *A. laeviceps*, both Asian species, the bivouac may be little more than a small, disk-shaped cluster of workers on the soil surface beneath the litter. The *A. laeviceps* cluster measures between 15 and 18 cm in diameter and 6 and 9 cm in height (Schneirla and Reyes 1966).

As exposed and ostensibly vulnerable as a bivouac can be, it nevertheless protects the brood from the capriciousness of the external environ-

Figure 3.30. An *Eciton hamatum* bivouac of the cylinder type, suspended from the trunk of a fallen tree. (Photograph by Carl W. Rettenmeyer, Connecticut Museum of Natural History.)

ment. Temperatures within the bivouac are significantly less variable than ambient conditions, and the constancy optimizes conditions for the developing brood (Schneirla et al. 1954, Jackson 1957). Temperatures within the bivouac of *Labidus praedator*, for example, vary much less than ground temperatures (Fowler 1979), and intrabivouac temperatures in *E. hamatum* are 1 to 2°C above ambient, and even higher in the compact brood mass. *E. hamatum* bivouac temperatures fall in early morning, rise in late morning, fall again in the afternoon, and rise in the evening despite fluctuations in ambient temperature (Jackson 1957).

At the core of the *E. burchelli* bivouac, the temperature is maintained at about 28.5°C, even though diurnal ambient temperatures vary only 6 or 7°C (Franks 1989b) (Figure 3.31). Franks (1989b) concluded from his investigation of intrabivouac temperatures that *E. burchelli* colonies actively engage in thermoregulation, and he hypothesized that high constant temperatures promote the high rate of brood growth and development typical of this and other epigaeic, phasic species. The heat required to maintain a constant bivouac temperature, Franks (1989b) proposed, could be generated by the basal metabolism of the ants in the bivouac. *E. burchelli* work-

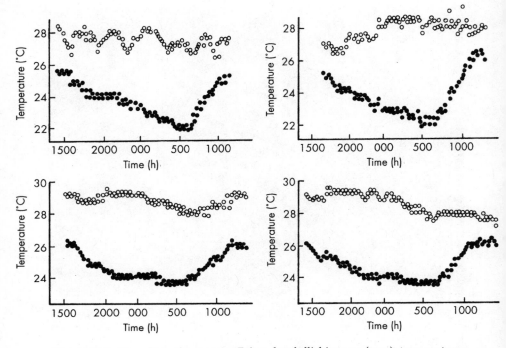

Figure 3.31. Circadian changes in *Eciton burchelli* bivouac (core) temperatures (open circles) and ambient temperatures (closed circles) for the same statary bivouac. Bivouac core temperatures are maintained at about 28.5°C, regardless of ambient conditions. (Redrawn, by permission of N. R. Franks and Blackwell Scientific Publications, from Franks 1989b.)

ers, for instance, produce as much as seven times the heat that they need to maintain a stable nest temperature. Temperature is regulated within the bivouac through the fashioning and closing of ventilation channels, much as occurs in bee swarms (Franks 1989b).

Like densely populated cities, army ant nests and their inhabitants produce garbage. And also like cities, ant colonies transport and segregate their garbage from the nest. Refuse deposits, some diffuse and others rather circumscribed, are probably formed by most army ant colonies a short distance from the nest or bivouac. Consisting of the remains of prey and the carcasses of dead workers, the refuse deposits quite naturally vary in content according to the diet of the colony. Little refuse can be found around *E. hamatum* bivouacs, for example, reflecting their diet of primarily soft-bodied prey (Rettenmeyer 1963b). On the other hand, the middens of *E. burchelli* are extensive and include the sclerotized fragments of consumed prey (they feed on a variety of adult arthropods), dead workers, and empty cocoons. Short "refuse columns" of workers extend outward from the bivouac to the deposits. Workers defecate in the refuse deposits and along the refuse trails—to such an extent that the feces may form distinct white paths during the statary phase (Rettenmeyer 1963b).

The middens commonly contain thousands of living arthropods, most of them scavengers on the discarded refuse (see Chapter 5). The fate of refuse in subterranean nests is all but impossible to determine, and references in the literature are rare. Carlos Bruch (1923) found refuse in three peripheral chambers of a leaf-cutter ant nest that had been occupied by a colony of *E. dulcius*. He noted that the refuse was replete with countless fragments—the heads and wings of other ants and the body parts of beetles and other arthropods, although it is quite possible that some of the discarded fragments were left by the original leaf-cutter ant occupants.

4

Behavior

It is the behavior of army ants, after all, that distinguishes them from all other ants. The organizational complexity of army ant colonies, as manifested in their behavior, is equaled or surpassed among the social insects perhaps only by the honey bee, the model social insect of entomological tradition, and by a few species of termites. Above all else, army ants are defined by the combination of group predation and nomadism they exhibit. As we have discovered, however, it is not the intrinsic nature of group predation and nomadism that makes army ants extraordinary, for both behaviors are found in other ants. It is, rather, the way these behaviors are immutably bonded in the ebb and flow of the colony's daily existence that makes army ants unique.

The reproductive biology of army ants also sets them apart from other ants, for like the honey bee, the army ant produces new colonies through colony fission. There is yet another distinction. Among the New World army ants and the Old World genus *Aenictus,* there are species that engage in a reproductive cycle of quiescent and dynamic periods that sometimes alternate with the precision of a chronometer—a timepiece whose mainspring is the maturing brood.

The integration of these behaviors into the group organization that constitutes the army ant adaptive syndrome depends on a communication system that is quite equal to the challenge. Individual social insects, as E. O. Wilson (1971a) noted, are not especially clever or complicated compared with individual solitary insects. What is novel about social insects, he pointed out, is their ability to turn simple individual patterns of behavior into mass phenomena by means of communication. Communication, therefore, assumes a pivotal enabling role in social organization. That is, the members of a society are like the cloth pieces of a patchwork quilt, communication being the thread that holds them together, juxtaposed in a pattern of social symmetry.

Functional Reproductive Cycle of the Colony

Army ant colonies that are functionally normal always contain developing brood. In some species, the brood are produced on a predictable, periodic schedule. Schneirla (1933, 1938) determined that the presence of larvae in colonies of these species somehow energizes the colonies to nomadic activity that involves daily or nightly emigrations to new nest sites. The presence of pupae results in depressed colony activity levels, the absence of emigration, and the initiation of the statary phase.

Based on his study of cyclic activities of colonies, Schneirla (1933, 1938, 1945, 1949, 1971) categorized army ants as belonging to either group A or group B species. Into group A, the phasic species, he placed the species that exhibit a well-defined cycle of alternating nomadic and statary phases that are conditioned by brood-stimulative factors (Figure 4.1). Several surface-active species of *Eciton, Neivamyrmex,* and *Aenictus* belong to this group. To group B, the nonphasic species, he consigned species that do not exhibit alternating phasic events. This group is characterized by variable nomadism—that is, emigrations to new nesting sites occur as single events separated by indeterminate intervals of non-nomadic behavior. The vast majority of army ant species appear to be nonphasic, which is probably the plesiomorphic, or less derived, condition.

The nomadic period of phasic species begins with the emergence of callow workers from their cocoons (Schneirla 1938). The nomadic phase is a time of elevated activity. For *Eciton hamatum* this means a large daily raid that begins at dawn and at dusk is converted into an emigration to a new bivouac site. Two brood stages are present in the colony at the very beginning of the nomadic phase: pupae that eclose as callow workers and young worker larvae hatched from eggs laid in the immediately preceding statary phase. During the nomadic phase, which lasts 16 to 18 days, the queen does not lay eggs and her gaster remains contracted. At first, the young larvae are maintained en masse at the center of the bivouac, which is generally exposed, but as they mature, the larger individuals are kept at the bivouac's periphery. The nomadic phase ends when the larvae spin their cocoons and pupate. As the larvae withdraw into their temporary shrouds, their stimulation of the workers ceases and the nomadic phase ends.

Figure 4.1. Functional reproductive cycle of phasic army ants. Eggs are deposited by the queen during the statary phase. Larvae hatch and develop during the latter half of the statary phase and during the entire nomadic phase. Pupation triggers the next statary phase. Emergence of callow workers stimulates the start of the next nomadic phase. Two cohorts of brood are present during the statary phase, one consisting of eggs and larvae, the other of pupae. During the statary phase, the queen is physogastric; in the nomadic phase, her abdomen is contracted.

The statary phase, which lasts about 18 to 21 days in *E. hamatum*, is clearly distinguished by subdued colony activity. Emigrations cease and foraging raids, although still carried out daily, are small and often weakly developed. The bivouac is established in a shelter such as a hollow tree or log, and the workers are less excitable and active. During the second week of the phase, the queen, now fully physogastric, delivers a new cluster of eggs, which constitute a unitary cohort. These hatch and larval development ensues. The statary phase ends when the pupae from the previous statary phase eclose from their cocoons as callow workers. Like the monotonous syncopation of a metronome, these phases alternate regularly throughout the life of a colony. Only when a sexual brood is present does the tempo alter.

E. *burchelli* has a similar cycle, although its nomadic phase of 11 to 16 days is considerably shorter and more variable than that of E. *hamatum* (Schneirla 1945). The statary phase in this species lasts 19 to 22 days.

An equivalent nomadic-statary cycle exists in *Neivamyrmex nigrescens*, although the functional cycle is completely interrupted during the winter months at the temperate extremes of its range (Schneirla 1958, 1961). In a further departure from the *Eciton* cycle, the first emigration of the nomadic phase following the eclosion of callow workers is not followed by emigration until about the fourth night of the phase. Additionally, the N. *nigrescens* colony remains in the nomadic phase until the larvae enter the early pupal stage. Schneirla (1958) postulated that nomadism in this species continues past the point of larval activity as a result of some stage-specific secretory or metabolic products or functions. It is possible that the stimulative cues of the larvae wane less abruptly than do those of *Eciton*, for unlike *Eciton* the pupae of *Neivamyrmex* do not spin cocoons; that is, there are no silken barriers between the pupae and the workers. The nomadic phase of this species in Arizona lasts for 20 to 31 days, the statary phase for about 18 days. During the statary phase, N. *nigrescens* workers become more photonegative and exhibit a stronger tendency to cluster, changes that correspond to the decrease in excitability of workers during this phase (Topoff 1975a).

According to Schneirla (1971), among the Old World army ants, *Aenictus gracilis* and A. *laeviceps* most closely resemble *Eciton* and *Neivamyrmex* in their functional cycle. As with *Neivamyrmex*, there are dramatic deviations from the *Eciton* prototype. During the nomadic phase, for instance, the *Aenictus* species are capable of foraging and emigrating at any time of day or night. When nomadic, their bivouac is a disk-shaped cluster of workers, either exposed or beneath litter, although near the end of the nomadic phase and throughout the statary phase the bivouacs are well sheltered or subterranean. Even though these two species have a nomadic phase of about 14 days and are thus similar to the New World phasic species, their 28-day statary phase is strikingly longer than that of the Ecitoninae. The statary phase begins when the advanced larvae enter a prepupal stage, but their emergence as callow adults is not associated with the beginning of a new nomadic phase, which instead begins some days later (Schneirla 1971).

Schneirla's exhaustive studies of phasic army ants were subsequently supplemented by numerous laboratory and field investigations. Studies of *E. burchelli,* for example, by Madalena Teles da Silva (1977a) confirmed the endogenous nature of the nomadic-statary cycles of this species. Extensive observations of *N. nigrescens* support Schneirla's theory that stimulation by the brood is a proximate cause of the nomadic period in phasic species; however, his theory does not account for characteristics such as frequency, direction, and distance of emigrations during the nomadic phase (Mirenda and Topoff 1980, Topoff et al. 1980a). Topoff and Mirenda (1980a) explored the relation between food supply and emigration frequency in *N. nigrescens* and concluded that hungry larvae stimulate relocation. Overfed laboratory colonies emigrated less frequently than colonies provided with little food (Topoff and Mirenda 1980a). In other words, army ants follow, in their own peculiar way, the dictum, attributed to Napoleon Bonaparte, that an army marches on its stomach.

It can only be assumed, given the paucity of data available on the biology of subterranean army ant species, that most army ants are non-phasic (Gotwald 1982). Even the surface-active species of *Dorylus* (*Anomma*), the driver ants, appear to be nonphasic. Certainly more is known about these conspicuous species than of any other species of Dorylinae. Although *D. (A.) wilverthi* emigrations are sometimes correlated with the eclosion of callow workers, emigrations are separated by intervals ranging from 6 to 40 days (Raignier and van Boven 1955). The East African driver ant, *D. (A.) molestus,* exhibits similar behavior. During 432 consecutive days of observation, one colony emigrated 38 times, staying at a single nest site for as little as 3 days and as long as 45 days (Gotwald and Cunningham–van Someren 1990). Raignier and van Boven (1955) recorded one intermigratory interval of 125 days in the West African species *D. (A.) nigricans.* It would appear that factors other than brood development serve as proximate cues for emigration in these species.

Even though the differences between phasic and nonphasic species are rather profound, Schneirla (1957) firmly believed that all army ants constitute a single monophyletic entity, and he attempted to homologize the functional cycle of *Eciton* with components of driver ant behavior. He concluded (1957, p. 294) that the first part of the *Eciton* nomadic phase and the single emigrations of *Anomma* involve "homologous reproductive processes as essential causes." In other words, the excitatory effect of eclosing callow workers is common to both. It then follows that each driver ant emigration should be regarded as a nomadic phase and the interval between emigrations as a statary phase. Schneirla (1957) was convinced that *Anomma* had a nomadic-statary functional cycle equivalent to that of *Eciton.* Given the evidence, however, it would seem that the alternating foraging and emigration patterns of these two groups are more analogous than homologous.

Communication

An activity on the part of one individual organism that alters the probable behavior of another individual organism in an adaptive manner constitutes biological communication (Wilson 1971a). Four sensory modalities are involved in insect communication: visual, vibrational, tactile, and chemical (Hölldobler 1984).

Most communication in social insects is based on chemical signals (Wilson 1971a), and this must certainly be true for visually limited species such as army ants. The vast majority of these chemical signals are in the form of pheromones, exocrine secretions of glandular origin released by one individual of a species that evoke a response in another individual of the same species. These chemical substances may have a releaser effect, in which specific, observable behavioral responses are evoked; or a primer effect, in which physiological alterations in the endocrine and reproductive systems result (Wilson 1971a). Pheromones assume added importance for species that characteristically maintain large colonies, since it appears that individual workers in larger colonies are more highly integrated into a webwork of chemical communication (Beckers et al. 1989).

Trail Pheromones

The nature of army ant locomotory behavior necessitates the effective use of odor trails. Whether foraging or emigrating, army ants are insects on the move and therefore require a means of coordinating groups and individuals into cohesive, mobile forces. Odor trails are an effective way of achieving that end (Blum 1974).

More than a century ago, naturalist Thomas Belt (1874, p. 22) observed the use of chemical trails by *Eciton hamatum* foragers:

Suddenly one ant left the conclave, and ran with great speed up the perpendicular face of the [tramway] cutting without stopping. It was followed by others, which, however, did not keep straight on like the first, but ran a short way, then returned, then again followed a little further than the first time. They were evidently scenting the trail of the pioneer, and making it permanently recognisable. These ants followed the exact line taken by the first one, although it was far out of sight. Wherever it had made a slight detour they did so likewise. I scraped with my knife a small portion of the clay on the trail, and the ants were completely at fault for a time which way to go. Those ascending and those descending stopped at the scraped portion, and made short circuits until they hit the scented trail again, when all their hesitation vanished, and they ran up and down it with the greatest confidence.

Although army ants' use of trail pheromones is indisputable, the source (or sources) of these substances has remained elusive. Watkins (1964) concluded that trail substances in *Neivamyrmex* may be in the feces, perhaps even added to the feces by some hitherto undescribed glands. Likewise, in *Eciton*, the hindgut was implicated as the source of trail pheromone (Blum and Portocarrero 1964). Ecitonines presumably lay their trails by dragging the tip of the gaster on the substrate surface. Both *Neivamyrmex* and *Eciton* have well-developed pygidial and post-pygidial glands with distinct reservoirs that open directly above the anal opening at the tip of the gaster (Hölldobler and Engel 1978).

Under experimental conditions, *E. hamatum* workers follow artificial trails drawn with crushed pygidial glands and prefer these trails to those drawn with hindgut contents or secretions from the poison gland and the Dufour's gland (Hölldobler and Engel 1978). But there may be multiple sources of trail substances, which further complicates the issue. For example, *E. burchelli* workers whose gasters have been removed and whose petioles are sealed with wax are still capable of laying a trail that elicits following behavior. The substance responsible for this redundant trail, perhaps a "footprint" secretion, is of short duration, indicating a volatile compound and suggesting therefore that it is probably not a true pheromone (Torgerson and Akre 1970b). Indeed, R. L. Torgerson and R. D. Akre (1970b) surmised that army ants will follow any substance that does not strongly repel them, and Howard Topoff and John Mirenda (1975) concluded that *N. nigrescens* can follow a variety of compounds that emanate from their body surfaces.

In laboratory investigations of trail following in *E. burchelli*, Johan Billen was able to identify the precise source of at least one trail pheromone. Extracts of the seventh abdominal sternite produced an impressive trail-following response in a worker exposed to an artificial trail (Billen 1992). Subsequent examination of the seventh sternite in *E. burchelli, Labidus praedator,* and *N. nigrescens* revealed a conspicuous glandular epithelium. Although the chemical nature of the pheromone was not determined, Billen (1992) characterized it as highly active and durable. A novel pair of reddish exocrine glands located beneath the seventh tergite of two species of *Aenictus* contain methyl-anthranilate and are a probable source of trail pheromone (Gobin et al. 1993).

The functional duration or stability of army ant trails, at least in *Eciton*, depends on four factors: (1) the nature of the substrate on which the trail is deposited, (2) whether the trail is made by a surface-active or subterranean species, (3) the type of trail—that is, whether it is a foraging or emigration trail, and (4) the amount of rainfall to which a trail is subjected (Torgerson and Akre 1970b).

In terms of substrate, trails laid down on porous substrates such as roots, logs, and lianas are more stable than those established on soil, leaf litter, and water pipes (Torgerson and Akre 1970b). Trails of epigaeic species such as *E. burchelli* and *E. hamatum* are more persistent than those of hypogaeic species. For instance, in an investigation conducted during the dry season, *E. burchelli* trails persisted for an average of 8.25 days, while a trail of the more subterranean *E. dulcius* lasted but 1.5 days. Emigration trails are more persistent than foraging trails because they are established and maintained by many more workers over a longer period. Finally, and not unexpectedly, trails on similar substrates are far more persistent in the dry season than in the rainy season (Torgerson and Akre 1970b).

As noted above, the chemistry of army ant trail substances remains to be elucidated. Murray S. Blum (1974), who has produced a significant corpus of work on exocrine secretions in social insects, observed that most trails are probably generated with mixtures of compounds. Because ecitonines produce trail pheromones that are active interspecifically, Torgerson and Akre (1970) proposed two components in trail pheromones: a general component that acts as a stimulus to all ecitonines and thereby serves as a general releaser, and a genus- or species-specific component that produces sustained trail following.

A lack of trail odor specificity is apparently common among the Ecitoninae. Watkins (1964) demonstrated that three species of *Neivamyrmex* (*carolinensis, opacithorax, nigrescens*) can follow one another's trails. In fact, two of the species followed trails prepared from extracts of whole workers of *Eciton dulcius*. This promiscuity in trail following did not extend to extracts made from non-army ants. In other trail preference tests in which *Neivamyrmex* species and *Labidus coecus* followed one another's trails, all species showed a distinct preference for trails made by conspecific workers (Watkins et al. 1967). This demonstrates that trail substances vary from species to species and that workers can detect the differences.

Certainly the chemistry of trail pheromones is complex. Although it is easy to demonstrate that army ants respond to relatively nonvolatile components in the odor trails, it has been shown experimentally that they also utilize volatile compounds (Topoff and Mirenda 1975). In *N. nigrescens*, these volatile substances have more than a single source: they are present in the deposited trail, and they are emitted directly by the ants from glands other than those that produce the deposited trail. Topoff and Mirenda (1975) speculated that because the distance between workers moving in narrow columns is small, the workers can respond to substances secreted from the body surfaces of trail-following nestmates.

Intraspecific subcaste and age differences in trail-following behavior also exist. For instance, although callow workers of *Eciton* have the same relative ability to detect and follow conspecific trails as do the mature workers, their running speed along the trails is significantly slower (Topoff et al. 1972b). Typically, callow workers do not participate in foraging excursions in the first few days after they eclose, and this delayed introduction to raiding activities may be adaptive. Perhaps their slower speed makes them less efficient predators than their speedier older sisters (Topoff et al. 1972b).

Several factors govern the involvement of callow workers in trail following. In *N. nigrescens,* callow workers do not take part in predatory raids until three to seven days following eclosion but do join their colony in emigrating to a new nest site within 24 hours of emergence. Curious about the recruitment of these callows to emigration but not to raiding, Topoff and Mirenda (1978) studied a colony in an observation nest. The newly eclosed callows formed a densely packed cluster surrounded by mature workers in the center of the nest. Prey collected on previous raids was placed at the interface of the callows and their mature sisters, as well as in piles elsewhere in the nest's periphery. The callows alternated between inactivity and feeding and grooming.

Topoff and Mirenda (1978) further discovered that as an evening raid progressed, the first workers to participate were those from the most peripheral areas of the nest. Foragers that had discovered prey returned to the nest and recruited additional workers, beginning with those at the periphery. In essence, the recruited workers were peeled away in layers from the central mass, but not enough ants were drawn from the nest to expose the core cluster of callows to the stimulatory effects of the aroused recruiters. Once emigration began, however, recruitment was sustained long enough to finally reach the callows, which then dispersed, eventually joining the exodus.

The major workers of *E. burchelli* and *E. hamatum* apparently are not as tightly bound to the chemical trail as are smaller workers. In other words, they are more readily distracted from the trail, especially by areas of disturbance. Under laboratory conditions, however, major workers responded to the trail substance with the same intensity as smaller workers did. Therefore, Topoff et al. (1973) concluded that the observed differences in responsiveness to the chemical trail in the field might be a result of major workers reacting to a combination of trail and alarm or

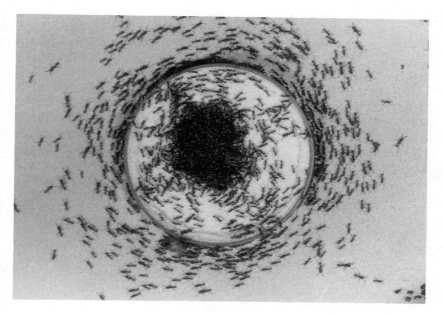

Figure 4.2. Under rare circumstances, army ants can be induced to follow a circular trail, shown here in *Neivamyrmex opacithorax*. T. C. Schneirla called this behavior circular-column milling. (Photograph by Carl W. Rettenmeyer, Connecticut Museum of Natural History. Reprinted, by permission of Carl W. Rettenmeyer and the University of Kansas, from Rettenmeyer 1963b.)

other pheromones. In the presence of alarm pheromone, for instance, these workers might be less responsive—less fettered—to the trail substance.

In general, however, army ants are all but shackled to their chemical trails, which they appear to follow slavishly. This is clearly illustrated in a phenomenon that Schneirla and Piel (1948, p. 22) designated circular-column milling. Under unique circumstances in nature (and rather ordinary conditions in the laboratory), army ants can be induced to form a tight circular column, a myrmecological merry-go-round, in which they "march themselves to death" (Schneirla and Piel 1948, p. 22). One such incident occurred among a group of about a thousand *Eciton* workers isolated on a sidewalk on Barro Colorado Island (Figure 4.2) (Schneirla and Piel 1948, p. 22).

They had apparently been caught by a cloudburst which washed away all traces of their colony trail. When first observed, most of the ants were gathered in a central cluster, with only a company or two plodding, counterclockwise, in a circle around the periphery. By noon all of the ants had joined the mill, which had now attained the diameter of a phonograph record and was rotating somewhat eccentrically at fair speed. At 10:00 p.m. the mill was found divided into two smaller counterclockwise spinning discs. At dawn the next day the scene of action was strewn with dead and dying Ecitons. A scant three dozen survivors were still trekking in a ragged circle. By 7:30, 24 hours after the mill was first observed, the various small myrmicine and dolichoderine ants of the neighborhood were busy carting away the corpses.

Schneirla and Piel (1948, p. 22) theorized that the circular track represented "the vector of the individual ant's centrifugal impulse to resume the march and the centripetal force of trophallaxis which binds it to its group." In a homogenous environment where there are no obstructions to disturb the geometry of these forces, the self-destructive mill is all but ordained. In nature, where the substrate is a heterogeneous collection of roots, stones, and leaves, the ant is not usually subject to this self-destructive behavior (Schneirla and Piel 1948).

A similar trail-following anomaly in *Eciton* was observed by William Beebe (1921, p. 289) at the Tropical Research Center of the New York Zoological Society in Kartabo, British Guiana (now Guyana). A bivouac had formed near the door of the outhouse, a most inconvenient place to impede the march of human progress. In an attempt to remove the inconvenience, three-quarters of the bivouac was destroyed with poison. The following morning, Beebe discovered that the remaining workers in the colony had formed an emigration column that issued from the outhouse, traversed a variety of substrates, including clumps of bamboo and underbrush, only to rejoin itself at its point of origin. The column was moving in a huge circle with a measured circumference of 1200 feet. To Beebe (1921, p. 291) it appeared as if "the ants fully believed they were on their way to a new home, for most were carrying eggs or larvae." He calculated that a worker could complete the round in about two and a half hours. The column continued for two days, with ever fewer workers marching as exhaustion took its toll; "burdens littered the line of march, like the arms and accoutrements thrown down by a retreating army" (Beebe 1921, p. 293). Finally, a few workers straggled from the circle, breaking the chain and leading the ragtag remnants of the colony off into the forest.

During the initial stages of raiding, before a strongly reinforced chemical trail has been established, tactile orientation may be of considerable importance to army ants. Raiding columns of *N. nigrescens*, for example, commonly run along the edges of rocks, logs, and fallen branches—

objects that may provide a tactile directional guide to the raiding front. In laboratory experiments with this species, Topoff and Lawson (1979) demonstrated that the responsiveness to tactile cues increases in the absence of a chemical trail and that there is a reciprocal and integrative relationship between chemical and tactile stimuli. This could be highly adaptive not only when a new trail is deposited but also when a trail is partially washed away by heavy rains. In the latter case, emigration columns may break up into small aggregations until the rain ceases, and physical objects along the trail probably facilitate reconnection and renewed movement of the column (Topoff and Lawson 1979).

Recruitment Pheromones

Certain specific social interactions have the effect of recruiting sister workers to tasks such as colony defense, nest construction, food retrieval, and emigration. In their most primitive form, these interactions consist of tactile signals; at their most derivative, however, chemical releasers become increasingly important. In army ants, chemicals appear to recruit workers to food, new terrain, and new nest sites. Although different chemicals may be employed in each behavior, the ants appear to use the same recruitment process in all three (Topoff 1984).

Army ants recruit sister workers to newly discovered food sources along chemical trails that are qualitatively different from foraging trails, a phenomenon first revealed in field studies of *E. burchelli* and *E. hamatum* (Chadab and Rettenmeyer 1975). After encountering a food source, the *Eciton* worker returns to the foraging column, intermittently dragging her gaster on the substrate as she goes. When the recruiter reaches the foraging column, she runs 5 to 10 cm in each direction in the column, contacting sister workers with her antennae and body. Periodically, she returns to and runs along the recruitment path, apparently reinforcing the trail by depositing additional pheromone.

Within 30 seconds, workers from the foraging column are diverted to the recruitment trail; in the first five minutes, 50 to 100 workers are recruited, drawn to the trail like iron filings to a magnet. Some of the recruited workers recruit new workers to the scene in a process that Ruth Chadab and Carl Rettenmeyer (1975, p. 1124) called secondary recruitment. Although there is a recruitment message in the trail substance, the effectiveness of the message is enhanced by the recruiter's behavior.

Field experiments disclosed that, in *Eciton* at least, (1) recruitment pheromone is either an entirely different substance from foraging trail substance or it is a combination of hindgut material and a secretion from some other source; (2) the army ant recruitment system, with respect to bringing together large numbers of workers quickly, is among the most efficacious in the ants; and (3) the efficient mustering of the large raiding forces that characterize army ant foraging is made possible by the concurrence of continuous foraging columns, the persistence of recruiters, a recruitment trail that both attracts and orients workers, and secondary recruitment (Chadab and Rettenmeyer 1975).

Laboratory experiments with *N. nigrescens* (Topoff et al. 1980b, Topoff 1984) confirmed the existence of a recruiting trail pheromone that is qualitatively different from foragers' exploratory trail substance. This recruitment pheromone alone is sufficient to initiate mass recruitment equal to that brought about by recruiters interacting tactually with nestmates. Topoff (1984) observed that recruitment to new terrain, as when a new foraging (exploratory) column emerges from the nest, and recruitment to food sources are similar processes, but that recruitment to new nest sites is a different matter. Recruitment to emigrate is unique because it involves the queen, callow workers, and brood-carrying workers.

During foraging, the queen, brood, and callow workers are sequestered near the center of the nest and are therefore insulated from the excitatory effects of returning foragers. When a new nest is located and emigration has begun, however, recruitment is continuous, and the queen and callow workers are eventually exposed to their aroused nestmates. They then join the decampment. But one question remains: What is the mechanism responsible for stimulating adult workers to pick up larval brood before joining the emigration column? Topoff (1984) hypothesized another pheromone (or pheromones) that stimulates broodcarrying behavior.

N. nigrescens workers that encounter food do not immediately return to the nest; first, they run several centimeters toward the nest, then turn and run toward the food. Topoff et al. (1980a, p. 785) called these back-and-forth runs looping movements. In the looping process, the highly excited workers encounter and arouse nestmates to follow the trail. Furthermore, such contacts stimulate these same nestmates to engage in the same looping movements characteristic of the primary recruiters. The secondary recruiters contact other workers in an ever-building crescendo of excitement. Indeed, it is secondary recruitment that initiates and sustains mass recruitment, and the bidirectional traffic of outbound and nestbound workers in the foraging column shifts to a unidirectional flow of outbound ants (Topoff et al. 1980a).

N. nigrescens preys on ant species that respond to attack by fleeing from their nests with their brood in tow, and the success of this army ant depends on its ability to maintain a ready attack force large enough for recruitment to all potential prey sites (Topoff et al. 1980b). This is accomplished through a number of strategies. First, mass recruitment (initiated by secondary recruitment) sets off a chain reaction that recruits thousands of workers to trail following. Second, mass recruitment can occur before prey is actually located; the principal source of arousal is simply the presence of new foraging ground. Recruitment to new foraging areas may be important in generating the multibranched trail system typical of this species, and which may be of singular importance in guaranteeing the presence of a large strike force within easy reach of any prey site discovered. When a termite or ant nest is found, mass recruitment to this food source takes precedence over recruitment to new foraging ground. Workers on trails near the newly found prey reverse themselves and stream toward the prey site. Third, the army ants tend to recruit more workers than are able to actually participate in the raid. This "recruitment overrun" (Topoff et al. 1980b, p. 787) does not produce confusion; rather, the superfluous workers eventually run past the raided nest and become part of a column that moves into new foraging ground.

Alarm Pheromones

Chemical alarm systems are common in the ants. More often than not, the pheromones that release alarm behavior are produced by the mandibular glands (Wilson 1971a). Although releasers of alarm behavior in army ants have not been chemically isolated or identified with any certainty, the existence of such substances is not in doubt. The dominant compound in the mandibular glands of *Eciton burchelli* and *Labidus praedator*, for instance, is 4-methyl-3-heptanone, but its role as a releaser of alarm behavior awaits experimental confirmation (Keegans et al. 1993). W. L. Brown, Jr. (1960) was the first to demonstrate that the alarm pheromone in *Eciton*, *Labidus*, and *Nomamyrmex* is produced in the head. The crushed head of a worker dropped into a column of workers elicited attack behavior during which the workers bit the head. On the other hand, crushed but headless bodies of workers attracted only momentary attention. In *Eciton* and *Nomamyrmex*, crushed worker heads emitted a meaty odor that Brown assumed to be associated with the alarm substance. Inanimate objects, such as twigs, rubbed against the crushed heads also elicited attack behavior when placed among the workers. Brown (1960) speculated that workers releasing alarm pheromone are not themselves attacked by sister workers either because they give off lesser amounts of alarm substance than are released by the crushed heads or because they also secrete, from the mesosoma or gaster, a substance that neutralizes the attack behavior of nestmates. The latter substance might be an identification pheromone or a nest odor.

Torgerson and Akre (1970b) repeated some of Brown's experiments to determine the extent of interspecific responses to ecitonine alarm pheromones. Although *E. hamatum* workers responded to crushed heads of *E. burchelli* with typical frenzied alarm behavior, the reaction was not reciprocal. *Nomamyrmex esenbecki* and *E. hamatum* responded to the other's alarm substance, but *E. burchelli* and *E. vagans* exposed to *N. esenbecki* alarm substance showed no response. Torgerson and Akre concluded that army ant alarm pheromones are more specific than is generally the case in the ants.

Queen and Colony Odor

Queen Odor

Ant queens produce pheromones with which they advertise themselves as reproductives. It has been empirically demonstrated in some species that workers learn to identify their queen's characteristic bouquet and can easily discriminate between their queen and an alien queen (Blum 1987, p. 287). It has also been shown that there may be a direct correlation between high pheromone production and the queen's fertility.

Army ant workers are attracted to their queen and recognize her odor. Although the source of the odor remains to be identified, the attraction is indisputable. In *Eciton*, for example, the queen is surrounded by a cluster of small workers in the undisturbed nest or bivouac and by larger workers when the nest is disturbed. In an emigration column she is always accompanied by a retinue of workers, which are typically attracted to the anterior portion of her gaster and rarely to her mouthparts or anal region (Schneirla 1949, 1971; Rettenmeyer 1963b; Rettenmeyer et al. 1978) (Figure 4.3). Under laboratory conditions, workers lick the queen's gaster more than any other part of her body (Rettenmeyer 1963b). Schneirla (1944) reported that workers are more attracted to the queen when she is physogastric (i.e., fertile) than when she is contracted, and this is in keeping with a correlation between high pheromone production and high fertility.

In laboratory experiments with *Labidus coecus* and five species of *Neivamyrmex*, Watkins and Cole (1966) found that although workers were

Figure 4.3. The queen of *Eciton rapax* with a small retinue of workers in an emigration column. (Photograph by Carl W. Rettenmeyer, Connecticut Museum of Natural History. Reprinted, by permission of Carl W. Rettenmeyer and the Connecticut Entomological Society, from Rettenmeyer 1974.)

attracted to the secretions of queens of other species, they showed a distinct preference for secretions of queens of their own species. Furthermore, the workers showed a clear preference for secretions of their own queens over those of queens from other conspecific colonies. Watkins and Cole also discovered that workers separated from their queen were more attracted to her secretions than were workers that were continuously housed with her. The secretions of queens remained effective in attracting workers for up to 72 hours.

Colony Odor

The ability to discriminate between nestmates and strangers is crucial to ants' social life. After all, distinguishing relatives from nonrelatives is essential to kin selection, and doing favors for nonrelatives is not adaptive. Sociobiology postulates "that altruistic behavior can evolve by natural selection only if the beneficiaries of the behavior are related to the donors [and] requires that helpers somehow direct their aid preferentially toward kin" (Carlin and Hölldobler 1983, p. 1027). Ants probably recognize their kin in one of two ways: either each colony member produces genetically determined odors or discriminators that, once distributed among all colony members, constitute a collective colony odor; or the queen produces the discriminators, which are subsequently dispersed among the colony's workers (Carlin and Hölldobler 1983). These discriminators then serve as recognition cues among nestmates.

Nestmates inspect one another with their antennae, using these sensory organs like paired divining rods that quickly identify the inspected individuals. If an intruder of a different species is discovered, it is violently attacked; if the intruder is conspecific, it may be treated less severely. Close-range olfaction appears to be the method ants use to identify nestmates and life stages. No colony odor by which nestmates can be recognized has been identified with certainty, but cuticular hydrocarbons may be the substances responsible (Blum 1987, Hölldobler and Wilson 1990). Even myrmecophiles appear to deliberately acquire their host colony's odor in the process of being accepted (see Chapter 5).

The existence of kin recognition chemicals is supported by anecdotal evidence. In my own field investigations, I have put these odors to practical use. If, for instance, I discovered two widely separated conspecific columns of driver ants and wanted to determine if they belonged to the same colony, I simply removed a worker from one column and placed it in the other. If this individual was accepted by the workers, I could assume that they were nestmates; if such a transfer resulted in hostilities, I judged that the columns originated from different colonies.

New World army ant colonies have a characteristic fecal odor that can be traced to the production of skatole (3-methylindole) by the workers (C. A. Brown et al. 1979). Although C. A. Brown et al. (1979) isolated the source of skatole as the head in *Neivamyrmex nigrescens*, Hölldobler and Engel (1978) proposed that it is also produced by the pygidial gland. Skatole and indole have been found in the gasters of *L. praedator*, and skatole has been discovered in the venom glands of some soldiers of *E. burchelli* (Keegans et al. 1993). The odor that W. L. Brown, Jr. (1960) associated with alarm pheromone (see above) in the ecitonines was more than likely skatole, and not the alarm substance itself. Skatole repels several species of insectivorous snakes (Watkins et al. 1969) and inhibits the growth of at least some bacteria and fungi (C. A. Brown et al. 1979). C. A. Brown et al. (1979) hypothesized that skatole is produced in the mandibular gland and is distributed over the brood and adults during grooming. They further proposed a dual function for skatole: it protects the colony from the harmful attacks of bacteria and fungi, and from predation by insectivorous snakes.

A

B

Plate 1. **A.** Linnaeus described the first army ant but assumed it to be a wasp and placed it in the genus *Vespa*. This is one of two *Dorylus helvolus* males from the Cape of Good Hope that Linnaeus had in his possession in 1764 when he described the species. His translated Latin description reads: wasp reddish brown, thorax hairy, feet rust-colored, femora compressed. (Photograph by W. H. Gotwald, Jr.) **B.** Emigration column of the ponerine ant *Leptogenys distinguenda* passing a group of guarding workers. This species clearly exhibits army ant behavior. (Photograph by D. Kovac, courtesy of U. Maschwitz.)

A

B

Plate 2. A. Workers of *Dorylus* (*Anomma*) *nigricans* returning to the nest with prey. Only a minority of workers carry visible prey items, but all presumably go back to the nest with liquids of prey origin stored in their crops. **B.** Caste-correlated tasks in *Dorylus* include an apparent defensive function for soldiers (majors). Here driver ant workers, *Dorylus* (*Anomma*) *gerstaeckeri*, assume a defensive posture near their colony's foraging column. (Photographs by W. H. Gotwald, Jr.)

A

B

Plate 3. A. An injured driver ant attacked by myrmicine ants. Army ant workers probably suffer considerable losses when foraging for prey. Wounded stragglers often fall prey to other ants, which easily overpower them. **B.** Foraging columns of some African species of *Aenictus* are weakly developed, consisting of small groups of three or four workers running together in single file. (Photographs by W. H. Gotwald, Jr.)

Plate 4. A curtain bivouac of *Eciton hamatum* between the buttressed roots of a tree. As exposed and ostensibly vulnerable as a bivouac seems, it nevertheless protects the brood from the capriciousness of the external environment. (Photograph by Roger D. Akre.)

A

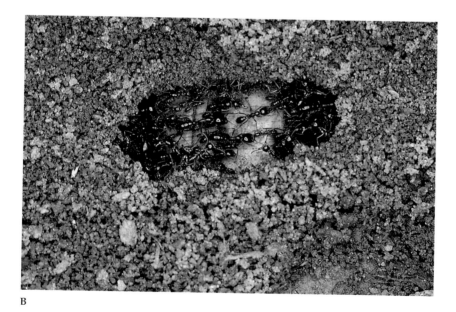

B

Plate 5. **A.** Driver ants commonly form soil particle walls and arcades that border and cover their foraging columns. **B.** The ants are visible through an opening in the arcade ceiling. (Photographs by W. H. Gotwald, Jr.)

A

B

Plate 6. Although driver ant foraging columns may be relatively small **(A)**, it is not unusual for raiding columns to achieve a density of 13 individuals per square centimeter **(B)**. (Photographs by W. H. Gotwald, Jr.)

A

B

Plate 7. A. Anastomosing columns of foraging driver ant workers close to the advancing swarm. **B.** Workers in the driver ant swarm as it crosses a road. The enormity of the swarm cannot be appreciated unless it is exposed on the substrate surface. (Photographs by W. H. Gotwald, Jr.)

A

B

Plate 8. A. Driver ants forage on low vegetation and drop back to the substrate when their search activities are concluded. Thousands of falling workers striking the forest floor produce the sound of a light rain shower. **B.** Driver ants, here *Dorylus (Anomma) gerstaeckeri*, commonly attack earthworms, which they simultaneously anchor in place and tear into transportable pieces. (Photographs by W. H. Gotwald, Jr.)

A

B

C

Plate 9. A stinkbug attempts to escape a driver ant attack by climbing vegetation **(A)**. Although the stinkbug releases a defensive secretion, the ants are not deterred **(B)** and soon overwhelm it **(C)**. (Photographs by W. H. Gotwald, Jr.)

A

B

Plate 10. A and **B.** Driver ant workers, like *Eciton* in the New World, cooperatively retrieve large or linear prey, such as these beetle larvae. (Photographs by W. H. Gotwald, Jr.)

A

B

Plate 11. **A** and **B.** All driver ant workers that retrieve prey, whether they do it cooperatively or alone, carry their prey slung beneath their bodies. (Photographs by W. H. Gotwald, Jr.)

Plate 12. A *Dorylus* (*Anomma*) *nigricans* trail bordered by soil particle walls where it crosses a road. At a distance, this engineering feat looks like an aerial view of the Great Wall of China. (Photograph by W. H. Gotwald, Jr.)

A

B

Plate 13. **A.** A closer view of the walled trail in Plate 12 reveals that the trail floor has been fashioned by the workers into a smooth, even surface. **B.** A *Dorylus* (*Anomma*) *nigricans* trail that has been excavated as well as having soil particle walls. (Photographs by W. H. Gotwald, Jr.)

A

B

Plate 14. A. Frequently, the walls bordering driver ant trails are made of the ants themselves. Many of the larger workers assume a defensive posture. (Photograph by W. H. Gotwald, Jr.) **B.** A carabid beetle, *Helluomophoides latitarsis*, eating larvae of *Neivamyrmex nigrescens* in the laboratory. These predators of army ant brood feed most intensively on nights when *N. nigrescens* colonies emigrate with their larvae. (Photograph by Howard Topoff.)

Plate 15. A major worker, or soldier, of *Dorylus* (*Anomma*) *nigricans* poised in the guard, or defensive, position. (Photograph by W. H. Gotwald, Jr.)

A

B

Plate 16. An emigration column of *Dorylus* (*Anomma*) *nigricans* with well-developed soil particle walls **(A)**. A closer view reveals that the workers carry their brood, like prey, slung beneath their bodies **(B)**. (Photographs by W. H. Gotwald, Jr.)

A

B

Plate 17. A. A white-throated antbird (*Gymnopithys salvini*), Tembopata Wildlife Reserve, Peru. (Photograph © Kenneth V. Rosenberg.) **B.** A swarm-following butterfly of the genus *Mechanitis* (family Nymphalidae, subfamily Ithomiinae). Neotropical butterflies, including ithomiids, feed on the fecal droppings of swarm-following birds. From these droppings the butterflies secure the nitrogenous nutrients necessary for egg production. (Photograph by Allen M. Young.)

A

B

Plate 18. A. Male army ants attracted to lights at night are often attacked by opportunistic predators, in this case a ponerine ant, *Megaponera foetens*. **B.** *Oecophylla longinoda* workers removing a driver ant worker from its foraging column. This is done so deftly that a widespread alarm response among the driver ant workers is avoided. (Photographs by W. H. Gotwald, Jr.)

Plate 19. A large spider with the *Dorylus* male it captured at a light. (Photograph by W. H. Gotwald, Jr.)

A

B

Plate 20. A. The territorial ant *Oecophylla smaragdina* (Asian weaver ant) attacking an *Aenictus* worker (in Malaysia). This ant is a formidable predator of *Aenictus* workers that violate its territory. **B.** Earthworms will actually climb vegetation to escape foraging driver ants. (Photographs by W. H. Gotwald, Jr.)

Plates 1A, 9C, 10B, 14A, 15, 16B reprinted from Gotwald 1984–1985, courtesy of the Royal Ontario Museum, Toronto, Canada. Plates 8A, 9A and B, 13B, 20B reprinted from Gotwald 1991, *Wildlife Conservation Magazine*, published by the Wildlife Conservation Society. Plate 14B reprinted from Topoff 1975b, courtesy of Howard Topoff.

Foraging

Hypogaeic and Epigaeic Lifeways

Most army ants are subterranean, or hypogaeic, foragers, or at least they forage concealed beneath leaf litter, logs, and rocks. Few venture to forage on the soil surface (i.e., epigaeically), where they are exposed to the dessicating effects of sun and air. Army ant species analyzed on the basis of where they forage and nest can be placed into one of three categories: (1) hypogaeic nesters and foragers, (2) hypogaeic nesters but epigaeic foragers, or (3) epigaeic nesters and foragers (Gotwald 1978b). It is safe to assume that most species that forage epigaeically also emigrate epigaeically. Still, in order to characterize the behavior of army ants accurately, the terms *hypogaeic* and *epigaeic* must be applied independently to three distinct components of army ant biology: nesting, foraging, and emigration.

Foraging (or raiding) schedules are not correlated with whether a species is an epigaeic or hypogaeic forager. Epigaeic *Eciton* species, for example, have a distinct diurnal routine, raiding from dawn to dusk, while *N. nigrescens,* also an epigaeic forager, raids from dusk until dawn. Epigaeic species of *Aenictus* initiate raids at all times of day or night (Schneirla and Reyes 1966). *Dorylus* (*Anomma*) *nigricans* and *D.* (*A.*) *wilverthi* also mount foraging expeditions day or night but show a preference for beginning in early evening and ending toward the middle of the following day (Raignier and van Boven 1955). Leroux (1975) found that 68 percent of the raids of *D.* (*A.*) *nigricans* that he observed in Ivory Coast began during the cooler hours, between 6:00 P.M. and 8:00 A.M. Hypogaeic foragers, such as *Labidus praedator* and *Nomamyrmex,* forage by day or night (Schneirla 1971).

Epigaeic foraging during daylight hours is not without its risks and constraints. Diurnal surface foragers may be sensitive to daily rhythms in temperature and may exhibit a midday lull in foraging between 11:00 A.M. and 2:00 P.M. that Schneirla (1949) termed the siesta effect. Common in *Eciton,* this period of diminished activity was also documented by Rettenmeyer (1963b), who noted that in *E. hamatum,* the foraging column might be reduced to one worker every 1 or 2 meters. He pointed out, however, that the column never became discontinuous. The construction of soil particle arcades over foraging columns by some species is probably an adaptive response to the physiological stresses of exposure. Such arcades are commonly constructed by African driver ants (Plate 5).

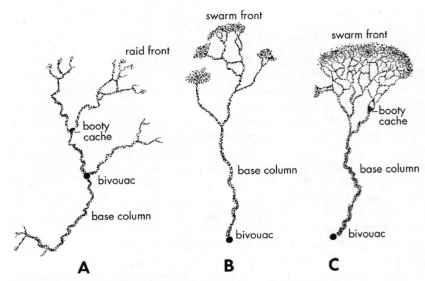

Figure 4.4. Two types of raiding patterns can be identified in army ants: column raids (A) and swarm raids (C), although intermediate patterns occur in some species (B). The column raid is typical of *Eciton hamatum*, the swarm of *E. burchelli*, and the intermediate type of *E. rapax*. (A and C redrawn, by permission of Carl W. Rettenmeyer and the University of Kansas, from Rettenmeyer 1963b. B redrawn, by permission of N. R. Franks and Blackwell Scientific Publications, from Burton and Franks 1985.)

Group Raiding for Prey

Foraging and Raiding Patterns

Schneirla (1934) detected two types of raiding patterns in army ants: column raids and swarm raids. The patterns are species specific, although gradations between the two are seen in species such as *Labidus coecus* (Rettenmeyer 1963b). As described in Chapter 1, a column raid consists of a system of branching columns of foraging workers diverging from a single base column that connects the nest with the foraging arena. Each branching column usually terminates in an advancing group of workers, which push forward into new territory with their antennae actively playing over the substrate and all objects before them (Figure 4.4). The direction that each terminal group takes is determined by the momentum of newly arrived workers and by topographical features. Membership in the advance guard changes constantly as workers push forward and then retreat, only to be replaced by others that behave in the same way (Schneirla 1934). In other words, there are no trailblazer specialists, just frenzied workers that pour from the advancing group like spilled water.

As raiding continues, trails to prey-depleted areas are abandoned. One advancing trail is usually maintained, however, and this gives rise to new branches as previously established trails are vacated. Thus, the one continuing base column grows progressively longer. The complex of branching columns is typically fan-shaped as it moves across the substrate (Schneirla 1934). Terminal raiding groups can advance rapidly— as much as 20 meters per hour in *E. hamatum*—and go considerable distances—as far as 350 meters in the same species (Schneirla 1971).

The complexity of column raids ranges between very simple, as exemplified by *Aenictus*, and quite complex, as seen in *E. hamatum*. Between these two organizational extremes lie species such as *Neivamyrmex nigrescens* and *Nomamyrmex esenbecki* (Schneirla 1971). In the Asian species *A. laeviceps*, the base column is one to five workers wide and often extends as far as 20 meters from the nest. The terminal groups range in width from a few centimeters to a few meters (Schneirla and Reyes 1966). Foraging columns in some African species of *Aenictus* are weakly developed, consisting of small groups of three or four workers running together in single file (Plate 3B). These groups are often widely separated, but all follow precisely the same trail, revealing its probable chemical basis (Gotwald 1976).

Neivamyrmex pilosus travels over its foraging trails in a similar manner, with gaps between individuals and groups (Rettenmeyer 1963b). *E. hamatum* raids usually develop three systems of trails on any one foraging episode, and the vigor of the raid varies with the functional cycle of the colony. The largest, most complex raids occur during the nomadic phase, when the workers are most excitable (Schneirla 1971).

A swarm raid is to a column raid what a six-lane highway is to a country road. The difference is dramatic. Certainly, it is the swarm raiding of army ants that has garnered the awed attention of naturalists and is responsible for the fictionalized image of voracious marauders that the very mention of army ants brings to mind. In a swarm raid, the base column divides in the foraging area into a series of anastomosing columns that coalesce to form a single advancing swarm of workers (Figure 4.4). As the swarm progresses across the substrate, it flushes all before it into panicked retreat. Invertebrates and vertebrates alike flee the onslaught; to stay is to become part of the army ant menu. It is to this phenomenon that the African driver ants owe their name. The Reverend Thomas S. Savage, who studied the habits of one West African species— probably *Dorylus* (*Anomma*) *nigricans*—observed in 1847 (p. 4) that "it not only travels and visits, in common with other species of ants, but it

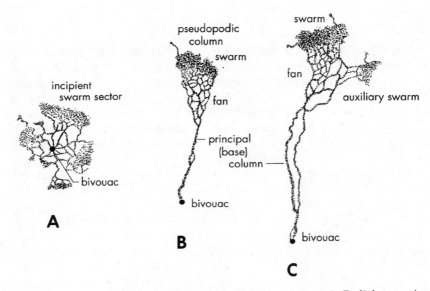

Figure 4.5. Progression of the *Eciton burchelli* raiding system. A. Radial expansion stage; the incipient swarm sector grows more rapidly than other sectors. B. Early development is characterized by establishment of the principal column, fan, and swarm. C. An auxiliary swarm is commonly produced by secondary division of the main body as the swarm advances. (Redrawn from Schneirla 1940.)

also *drives* every thing before it capable of muscular motion, so formidable is it from its numbers and bite; in respect to the last fact it stands unique in its habits, and, in distinction from other species of this country, may well take for its vulgar name that of *Driver*." *E. burchelli* swarms are commonly more than 5 meters wide and occasionally surpass 15 meters (Schneirla 1971), while in some species of driver ants swarms are likely to be 20 meters or more across (D. H. Kistner and W. H. Gotwald, Jr., unpubl. data).

Schneirla (1940) described the principal features of the developing swarm raid in *E. burchelli*, which develops anew each day. At daybreak, the increasing light intensity, a solar alarm clock of sorts, arouses the workers. (The role of light intensity was speculation on Schneirla's part. The alternative possibility of an internal alarm clock—that is, an intrinsic circadian rhythm—should not be overlooked.) Some of the aroused workers move from the bivouac to the ground, then begin advancing from the bivouac, hesitating and retreating repeatedly, expanding in all directions. This part of swarm raid development Schneirla (1940, p. 412) termed the period of radial expansion. Soon, one sector of this perinidal

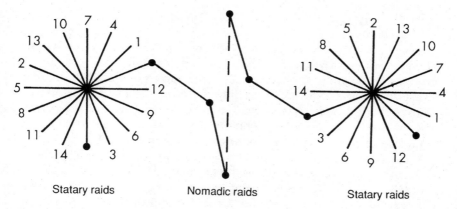

Statary raids   **Nomadic raids**   **Statary raids**

Figure 4.6. The behavioral cycle of *Eciton burchelli*. Each statary phase is represented by raids that systematically radiate out from the bivouac, thereby reducing foraging overlap. Numbers indicate sequence of raids. (Redrawn, by permission of N. R. Franks and *American Scientist*, journal of Sigma Xi, the Scientific Research Society, from Franks 1989.)

area becomes favored over the others, and workers are drawn to it from all sectors (Figure 4.5). This congregation of workers, which assumes an elliptical form, constitutes the incipient swarm. As the swarm sets out, it is connected to the bivouac by a triangular-shaped series of anastomosing columns that Schneirla (1940, p. 414) dubbed the fan. But eventually these feeder columns drop away, leaving the bivouac and foraging swarm connected by a single principal (base or trunk) column (Schneirla 1940, p. 414) (Figure 4.5).

Schneirla (1940) attributed the direction of swarm movement, at least in part, to the greater exodus of ants from the side of the bivouac facing the favored sector. He did not, however, notice any pattern in the directions chosen by swarms on successive mornings. Franks and Fletcher (1983), on the other hand, found that successive statary phase raids of *E. burchelli* rotate about the bivouac site systematically so that foraging overlap is minimized. The mean angle between successive raids is 123°, a pattern analogous to the spiral arrangement of leaves (spiral phyllotaxis) on some plants that minimizes self-shading, although plant spirals are more concise (Franks and Fletcher 1983). Army ant raids resemble rather thin, long leaves that do not require precise arrangement to avoid overlap. Franks and Fletcher also found evidence of systematic navigation in raids conducted during the nomadic phase that resulted in greater separation of successive statary bivouacs than could be achieved randomly (Franks and Fletcher 1983) (Figure 4.6).

As the morning unfolds, the *E. burchelli* raiding swarm advances 90 meters or more from the bivouac. Early on, the swarm moves at a relatively constant rate of 12 to 14 meters per hour, and as it moves it grows to a width of 12 meters or more. Later in the morning, the swarm numbers may reach unmanageable proportions, and it may divide to produce an auxiliary mass of workers that diverges from the original swarm (Schneirla 1940). At its zenith, the *E. burchelli* swarm is an immense elliptical body of ants that contains more than 25,000 workers and may be 15 meters wide (Schneirla 1940). All this time, the swarm maintains its general direction of advance with few deviations.

There are no leaders as the swarm moves forward, since trail pushing is done by any and all raiders that enter new ground (Schneirla 1971). Schneirla (1940, p. 421) termed this exploratory behavior of advancing workers the rebound pattern and divided it into three components. In the first, the track phase, a worker moves with little hesitation toward the front of the swarm over the presumably saturated chemical trail established previously by her colony mates. In the second, the pioneering phase, the advancing worker reaches the chemically unmarked terrain at the forward border of the swarm. Her advance is abruptly interrupted by the sudden lack of stimulation, and she "recoils back into the mass" (Schneirla 1940, p. 422). This is followed by the retreat phase, the third and final component, in which the worker passes back over chemically saturated ground, encountering as she goes her advancing nestmates. She then reverses or alters her direction and again flows with the swarm.

Organization of the swarm, Schneirla (1940, p. 426) hypothesized, depends on the positive pressure exerted by the steady stream of outbound workers arriving at the swarm and by an "impedance" of the swarm by the slower pioneer foragers at the front and by "local stoppages in pillaging operations" (Figure 4.7). The impedance compels newly arriving workers to move laterally, thereby widening the swarm. The opposition of these two forces creates the elliptical shape typical of the swarm. Pressure on the swarm from workers deploying laterally to the swarm's flanks is equibalanced. In this way, the swarm keeps to the same direction and faithfully advances away from the bivouac (Schneirla 1940).

Although the *E. burchelli* swarm maintains a given direction of progress, it does so by alternating slow flanking movements (Schneirla 1940). Like an ever-extending pendulum, the entire swarm turns first to one side and then to the other in swings of 15° to 25°. The flanking movements give the swarm a meandering course that is reflected in the principal trail's serpentine appearance (Schneirla 1934). As a flanking movement occurs, the workers become concentrated within the side of the swarm favoring the flanking direction (Figure 4.7). In the opposite

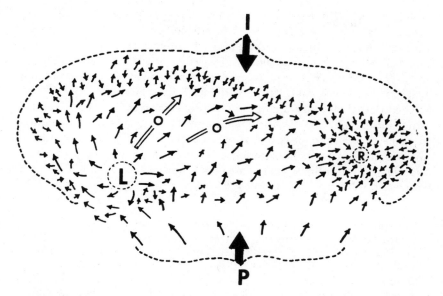

Figure 4.7. A flanking movement in an *Eciton burchelli* swarm. Organization of the swarm depends on pressure (P) exerted by arriving workers and an impedance (I) to mass movement caused by the slower advance of pioneer foragers at the swarm front. In this example the right flank becomes concentrated (R) while the left flank undergoes expansion (L). (Redrawn from Schneirla 1940.)

wing of the swarm, the workers spread forward and obliquely inward. Thus the entire swarm appears to pivot on one wing, like a gate on animated hinges. Schneirla (1971) considered these flanking movements superior in organization to the advances of other swarm raiders and believed that they permit *E. burchelli* to hold a single direction of advance better than the swarms of *Labidus praedator* or the driver ants.

The foraging swarm thus represents a self-organizing collective body whose movements are more cohesive than could be anticipated in the absence of any centralized control. J. L. Deneubourg and colleagues developed computer models that demonstrate how swarms can be self-generating. Their models, based on Monte Carlo simulations that took into consideration the interactions of individual foragers and their ability to lay and follow chemical trails, produced advancing swarms that were "dynamic and adapt as the actors interact with the environment, conferring a degree of intelligence to the society that far exceeds the capacity of its individual members" (Deneubourg et al. 1989, p. 724).

The principal trail that supplies the *E. burchelli* swarm with workers is qualitatively different from the numerous defunct branch trails abandoned as the swarm advances, a difference Schneirla (1940) attributed to its chemical distinctness compared with lesser-used trails. This sinuous trail is occupied by a column of workers about 3 cm wide. Early in the day, traffic on the trail is predominantly away from the bivouac. Later, movement is much more variable (Schneirla 1940).

The internal organization of the foraging swarm of African driver ants is not unlike that of *E. burchelli*, although the swarm is more grandiose in scale and perhaps less orderly (Schneirla 1971). Observing a driver ant swarm is a stunning experience, even for the most jaded of travel-experienced naturalists. One of my own journal entries describes these foragers on the move (Gotwald 1984–1985, p. 37):

> This morning I watched a colony of driver ants forage. The advancing swarm of worker ants moved with the effortlessness of a rain-swollen river. It flowed across the forest floor with singleness of purpose, altered in its course only by the most intrusive of natural barriers. At the forefront of the swarm, hearty explorers reached out, as if they were the fingers of this fleeting hand, determining in some instinctive way the path to be followed. Behind the swarm, orderly columns of moving workers created a sense of organization for what superficially appeared to be monumentally chaotic. These columns merged to give the swarm its body. As the swarm progressed over the forest litter, small stationary groups of workers formed, giving the moving mass of foragers the appearance of an island-choked delta. In advance of the swarm, all manner of creatures fled, warned in some mysterious way of impending danger. It would be fatal to face the onslaught.

The raid begins most frequently at dusk in *D. (A.) wilverthi*, less often at dawn, and perhaps never at midday. The direction taken by the incipient swarm appears to be a matter of chance and depends on the nest opening from which the workers first emerge (Raignier and van Boven 1955). These workers form a rather disorderly mass, out of which orderliness arises in the form of several small anastomosing columns of ants (Figure 4.8). Near the nest, these columns coalesce to form the principal, or trunk, column (Raignier and van Boven 1955).

D. (A.) nigricans raids have a somewhat different beginning. According to Leroux (1977b), raids begin with larval-induced excitement in workers. Initially, the foraging workers spread out from the nest, sheetlike, until a fan-shaped network of columns is formed. Seldom does a raid begin without the formation of this sheet (*nappe*) of workers (Leroux 1977b, p. 447). One gets the impression from Leroux's (1977b) descrip-

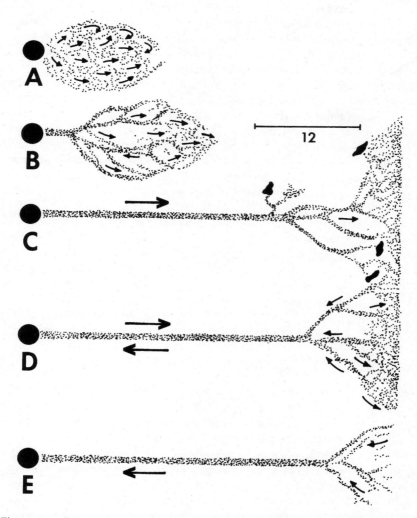

Figure 4.8. Formation and demise of the *Dorylus* (*Anomma*) *wilverthi* swarm. Growth of the swarm includes an initial dispersion (A), organization of anastomosing trails and the principal trail (B), outward movement of workers to the foraging arena (C), bidirectional movement of workers on the principal trail as foragers return with prey (D), and termination of the raid (E). Scale in meters. (Redrawn, by permission of A. Raignier and the Africa Museum, Tervuren, from Raignier and van Boven 1955.)

tion that the *nigricans* raid moves at first like a giant amoeba through advancing pseudopodia formed of energized workers.

The driver ant raiding swarm advances as rapidly as 20 meters or more per hour and deviates from its general direction at greater angles than does the *E. burchelli* swarm. Raiding columns become so crowded with workers that they may achieve a density of 13 individuals per square centimeter (Raignier and van Boven 1955) (Plate 6). The average length of the trunk column in *D. (A.) wilverthi* and *D. (A.) nigricans* is 125 meters according to Raignier and van Boven (1955), although 63 percent of the *nigricans* raids observed by Leroux (1975) measured between 26 and 75 meters. Raiding expeditions in driver ants commonly last from 9 to 27 hours (Raignier and van Boven 1955: Leroux 1975, 1977b) (Plate 7).

A comparison of column and swarm raiding reveals two strategies: in the former, a series of terminal raiding groups forages over narrow strips of substrate; in the latter, a single mass of workers sweeps across a wide area of substrate. It would seem that the latter strategy is the most adaptive for the general predator because it provides a mechanism for encountering simultaneously a diverse group of prey. The former strategy, on the other hand, apparently functions well for species that are oligophagous or that prey exclusively on other social insect colonies.

Group Retrieval of Prey

In both column raiding and swarm raiding, foraging workers respond to the movements of potential prey, even though some workers (the Old World army ants) are blind. Perhaps blind or nearly blind workers detect prey through substrate vibrations. Prey odors are surely important in prey detection, too, although evidence for the use of olfactory cues by foraging workers is lacking. Column raiders are not as responsive to prey movements as swarm raiders, and they respond first to prey odors (Schneirla 1971). The swarm raiding *E. burchelli* observed by Rettenmeyer (1963b) were so responsive to movement that raiding workers would even attack moving blades of grass. Some arthropods apparently avoid attack, even when surrounded by foragers, by remaining motionless (see Chapter 6).

When searching for prey, *E. hamatum* workers run primarily on the ground and on leaf surfaces. Although they do climb vegetation, they rarely go above 2 meters (Rettenmeyer 1963b). This contrasts markedly with some driver ant species, which forage in trees to heights of more than 3 meters (Raignier and van Boven 1955, Gotwald 1972a), and with *E. burchelli*, which raids into the treetops (Rettenmeyer 1963b). When foraging on low vegetation, a meter or so tall (Plate 8A), driver ants habitually drop back to the substrate at the conclusion of their search activities rather than climbing down. Thousands of falling workers striking the forest floor produces the sound of a light rain shower. Many times in the field I have discovered foraging driver ants by listening for this telltale sound of rain. Dropping to the substrate most likely reduces the energy costs of search activities. As with most epigaeic foragers, the search behavior of some African *Aenictus* species involves a meandering exploration of the soil surface and crevices and holes in the soil (Gotwald 1976).

Prey captured by foraging workers are immobilized, dismembered, often sectioned into transportable pieces, conveyed to the nest, and distributed to nestmates. Foraging workers bite their prey and in some instances also sting them. Large prey such as scorpions that strongly resist attack are pinned down by workers anchored by their tarsal claws, spread-eagled by oppositely pulling groups, and torn apart (Schneirla 1971). Earthworms are attacked in this manner by *Anomma* driver ants. While some workers anchor the writhing earthworm in place (a decidedly difficult task), others tear small pieces of tissue from the captive (Gotwald 1974b) (Plate 8B). New World species both bite and sting their prey, although the swarm raiders *E. burchelli* and *Labidus praedator* are considered more potent than column raiders in both respects (Schneirla 1971). Even though *Dorylus* workers possess morphologically complete stings, they do not use them, instead relying exclusively on biting their prey (Gotwald 1978) (Plate 9). Their sharp, cutting mandibles, plus their great numbers, permit *Dorylus* to kill and dissect even vertebrates, prey animals not commonly attacked by New World species (Schneirla 1971). The ability to sting is variable in *Aenictus*. The Asian *Aenictus* (*gracilis* and *laeviceps*) observed by Schneirla (1971) possess potent stings and strong bites, while African species do not sting (Gotwald 1978b).

Large prey organisms are usually sectioned before being transported to the nest, but even the smallest of prey may have its appendages torn from it. Linear prey units such as centipedes and beetle larvae are usually carried by two or more cooperating workers running in tandem (Plate 10). Other prey items are carried by individual workers (Plates 2A, 11). All retrievers of prey, whether individuals or cooperating workers, straddle their prey and carry it slung beneath their bodies. Franks (1986, p. 425) discovered that cooperating workers of *E. burchelli* form teams that exhibit a distinct caste distribution and are "superefficient" in retrieving prey (see Energetics of Foraging, below).

William Beebe (1921, p. 74), whose purple prose enshrined army ants as nature's quintessential predators, described the return to the bivouac of booty-laden *Eciton* foragers: "Scorpions, eggs, caterpillars, glass-like wasp pupae, roaches, spiders, crickets—all were drawn into the nest by a maelstrom of hunger, funneling into the narrow tunnel; while from all the surface of the swarm [bivouac] there crept forth layer after layer of invigorated, implacable seekers after food." Initially, raiding traffic on the principal and branch trails is unidirectional away from the nest. But as foragers encounter, capture, and return with prey, traffic on the trails becomes bidirectional. Finally, as the raid loses momentum and comes to a close, traffic on the trails becomes undirectional again, now toward the nest. This temporal shift in traffic has been observed in *Eciton* (Schneirla 1971), *Dorylus* (*Anomma*) (Raignier and van Boven 1955), and *Aenictus* (Gotwald 1976). In highly productive raids, especially by *Eciton* and *Aenictus,* large caches of prey are sometimes deposited at points where columns branch from one another near the foraging area. This prey is later retrieved and returned to the nest (Rettenmeyer 1963b, Schneirla and Reyes 1966, Schneirla 1971) (Figure 4.4).

Driver ants and *Labidus* commonly form walls and arcades, either of soil particles or of the ants themselves, to border and cover their foraging columns (Plates 5, 12, 13, 14A) (Cohic 1948, Raignier and van Boven 1955, Rettenmeyer 1963b). This phenomenon may occur to an even greater extent along emigration trails (Plate 16) (Schneirla 1971). The walls and arcades probably provide shade and increased humidity, especially important for columns that must cross exposed surfaces. Furthermore, the walls of clustered workers may serve to keep itinerant arthropods and other small invertebrates from accidently straying into and disrupting the flow of traffic. I have witnessed *Anomma* workers repel insects in this way. Major workers positioned at trail margins often assume a defensive or guard posture with heads elevated and mandibles held wide apart (Plates 2B, 15) (Gotwald 1982).

The number of foraging workers that return to the nest with visible pieces of prey is small in proportion to the total number of individuals that participate in a raid (see discussion of polyethism in Chapter 3). What is certain in driver ants, and probably is true of other army ants as well, is that smaller workers not laden with pieces of prey are transporting prey liquids stored in their crops (see Figure 3.11).

Energetics of Foraging

Like other ants, army ants are central place foragers. That is, they do not consume their prey where it is captured but rather return it to a central place, the nest, where it is eaten, stored, or fed to the larvae. General foraging theory assumes that a forager increases its fitness by maximizing the rate of delivery of energy to the central place (Orians and Pearson 1979). The basic unit in central place foraging theory is the round trip, composed of an outward excursion, a foraging period, and the return trip. Although, as Gordon Orians and N. E. Pearson (1979) noted, energy is utilized in all three phases of the trip, it is obtained only during the second phase. They further pointed out that the return trip is more expensive per unit distance traveled than the outbound trip because of the added cost of carrying prey. Analyses of the energetics of army ant foraging should reveal the level of foraging efficiency achieved by these ants and the selective advantage that accrues to group predators.

Franks (1985, p. 91) concluded that the division of labor among *E. burchelli* workers may increase foraging efficiency, and that this, in turn, should increase the colony's growth rate, decrease its generation time, and contribute to the inclusive fitness of all members of the colony. Among the four demonstrable worker subcastes of this species, submajors appear to specialize as porters hauling prey items from the raiding front to the bivouac. Proof of their occupational specialty, Franks (1985) noted, is apparent in the fact that 26 percent of the workers carrying prey are submajors, although they represent but 3 percent of the worker population.

Submajors are well adapted to their job. First, they have longer legs in proportion to their body size than any of their sister workers and when not burdened with prey can run faster than any of their colony mates. Second, although all workers run at the same speed when carrying prey, submajors can carry disproportionately larger prey items. Indeed, they transport the greatest weight per unit of their own weight of any worker subcaste. Both adaptations may contribute to reducing transport costs, since workers must be able to deliver their prey to the nest quickly and return to the raiding front to capture and retrieve even more prey. The immensity of the task is made clear when we realize that 30,000 prey items are returned to the bivouac over a distance of 105 meters during an average raid (Franks 1985).

The porter subcaste of *E. burchelli* fits predictive models for efficient transporter design. Franks (1985) noted, for instance, that the net cost of transport—that is, the amount of energy expended to transport one unit of body weight over one unit of distance—is inversely proportional to body weight. In other words, the unit cost of transport diminishes as the body mass of the transporter increases. In reality, the submajors are the largest workers carrying prey—the ice tong–like mandibles of the majors (or soldiers) disqualify them as porters. Franks (1985) also noted that the speed of workers is probably a positive function of leg length. The submajors, with their disproportionately longer legs, probably transport prey faster at a relatively lower energetic cost.

It is quite probable, then, that polymorphism and a correlated division of labor reduce foraging costs. But what of monomorphic army ant species? How do they minimize prey transport costs? Curiously, the genus *Eciton*, noted for the extreme polymorphism of some of its species (see Chapter 3), also includes the essentially monomorphic species *E. rapax*. A specialist predator on other ant species, *E. rapax* conducts raids over long distances, frequently greater than 250 meters. Although the species exhibits considerable variation in worker size, the workers' body proportions scale isometrically. In other words, allometric variation between workers and distinct subcastes is absent (Burton and Franks 1985). Although there are no soldiers, *E. rapax* workers are the largest of any *Eciton* species (Rettenmeyer et al. 1978). *E. rapax* raids in many small but linked swarms, a pattern intermediate between the swarm-raiding pattern of *E. burchelli* and the column-raiding pattern of *E. hamatum* (Figure 4.4). This species provides an unusual opportunity to compare the foraging efficiency of a monomorphic species with that of its polymorphic relatives (Burton and Franks 1985).

Their analysis of the foraging ecology of *E. rapax* led J. L. Burton and Nigel Franks (1985) to conclude that the role of colony defense is rather evenly distributed among the *E. rapax* workers. They are endowed with a potent sting, perhaps the most painful among the Ecitoninae, and their black alitrunk and bright yellow gaster suggest they are aposematic— that is, they exhibit a distinct warning coloration pattern (Rettenmeyer 1974). Burton and Franks further concluded that all workers are forced to participate in transport of prey and brood (when emigrating) over exceptionally long distances. Finally, transport costs have selected for monomorphism in this species. Because these costs are a large fraction of the colony's energy expenditure (workers travel long distances when raiding and emigrating because they prey on ants that nest in widely dispersed colonies), large workers that can carry disproportionately large loads should be selected for. The unit cost, remember, declines as the size of the carrier increases.

E. burchelli and *E. hamatum*, the two most strongly polymorphic *Eciton* species, may have been able to reduce their foraging distances by broadening their diet, although they also prey on the same species as *rapax* (as revealed in Table 4.1, *hamatum* is much more the specialist than *burchelli*). This, in turn, made it possible for them to reduce their transport costs by using a specialized, small fraction of their workers as porters.

Consequently, the average size of workers in these two species is smaller (Burton and Franks 1985).

Observing that *E. burchelli* workers often formed groups to cooperatively retrieve large linear prey items, Franks (1986) investigated the membership structure of these groups. His research revealed that cooperative workers function as teams. He discovered, for instance, a constant relationship between the dry weight of prey carriers and the weight of the prey they carry and concluded that workers are somehow able to evaluate their own performance and their potential contribution to the collective effort. A large prey item is never picked up simultaneously by a group of collaborating workers. Instead, the group forms gradually. The prey item is first moved by a large worker, usually a submajor of the porter caste. Additional smaller workers join in the effort until the prey item is being transported at the standard retrieval speed. Once this speed is achieved, no more workers join the group (Franks 1986).

The fact that groups begin with a single large worker explains why groups most commonly include one submajor, although other combinations of workers are possible. Because unit transportation costs decrease with increasing worker size, it should not be surprising that a large prey item is always carried by a submajor first, if one is available, rather than by two smaller individuals.

When Franks (1986, p. 428), called *E. burchelli* teams superefficient, he meant that if a prey item being carried by a team were divided into pieces, one for each team member, the members would be unable to carry their assigned individual pieces. In theory, two workers cooperating in prey retrieval would be expected to carry as much as three times what one worker could carry (Franks 1986). Furthermore, hypothetical dictates of efficiency require that teams should optimally consist of only two workers, since there is little advantage gained in adding more individuals (Franks 1986). This supposition relates to the number of worker legs that can be kept on the ground per capita during running. There is not a proportional increase of such legs on the substrate when individuals are added to the team. The hypothetical appears to be confirmed: 88 percent of the *E. burchelli* teams Franks observed consisted of only two workers (Franks 1986). Similarly, driver ant teams commonly include but two members (Plate 10).

The energetic performance of *E. hamatum* was investigated and compared with that of the leaf-cutting ant *Atta colombica* by G. A. Bartholomew et al. (1988) and D. H. Feener et al. (1988). Leaf-cutters are fungus-culturing myrmicine ants that belong to the tribe Attini. They commonly forage for fresh vegetation, which they carry back to the nest and process into a substrate on which they grow fungus, which they harvest and eat (see Hölldobler and Wilson 1990). Bartholomew et al. (1988) determined the following about *E. hamatum:* (1) the running speed of minor workers increases linearly with body mass, and major workers travel at about the same speed as large minor workers; (2) the running speed of unladen workers does not differ significantly from that of laden workers; (3) oxygen consumption of running workers increases allometrically with body mass, and the energy cost of unit transport decreases as running speed increases; (4) the gross cost of transport decreases with increases in both body mass and running speed (speed being the more important); and (5) the energy cost of carrying a load does not differ from carrying the same mass as a part of the body, so the energy costs of transport of either laden or unladen ants can be calculated (at a given temperature) if running speed, body mass, and load mass are known.

Feener et al. (1988:517) concluded that

> In comparison with *A. colombica, E. hamatum* is smaller, runs faster, harvests foodstuffs richer in usable energy and carries lighter loads both absolutely and relative to its mass. Parallel to these differences, per capita rates of energy expenditure are lower in *E. hamatum* for both unladen workers and workers carrying loads of a given load ratio. The lower cost of load carriage in *E. hamatum* suggests that the army ant mode of prey carriage with the load slung under the body and between the legs is mechanically more effective than the overhead load carriage used by leaf-cutters. However, as load ratio increases, the energy costs of load carriage relative to the energy costs at rest increase more rapidly in *E. hamatum* than in *A. colombica.* Thus, while the army ant mode of load carriage may be mechanically more effective, leaf-cutters are better adapted for carrying heavy loads.

The foraging energetics of Old World species has not been investigated. Given the size and scope of driver ant colonies and their foraging expeditions, they beg the same attention given to *E. burchelli* and *E. hamatum. Aenictus,* too, as small monomorphic ants, merit consideration. Until their foraging efficiency is analyzed, our understanding of group predation remains incomplete.

Trophic Relationships within the Nest

The movement of food within the army ant nest is not well understood. Obviously, the trophic focus of each colony is the larvae. Indeed, the larvae's state of satiety or hunger has a pronounced effect on the foraging and emigration behavior of the colony (Topoff 1984). Larvae probably feed on prey solids provided by the workers, either pieces or whole prey items. As I noted in Chapter 3, W. M. Wheeler and Bailey (1925) suspected that *E. burchelli* larvae are fed at considerable intervals with large food pellets composed of the masticated, rolled-up soft parts of prey. Food pellets are formed in the worker's infrabuccal pocket, a diverticulum of the posterior hypopharynx that, in some species, filters solid fragments from fluids that are to be stored in the crop (see Gotwald 1969).

As army ant larvae grow larger, they can feed directly on whole pieces of prey. Schneirla (1971) noted that the larvae use movements and odors to attract the workers that feed them, and that as larvae mature, *Eciton* workers become increasingly attentive and more frequently drop food items on them. Older larvae are also more often carried by workers and dropped on caches of prey in the nest (Schneirla 1971). Even though foraging workers of *Anomma* driver ants return to the nest with prey liquids stored in their crops, they have not been observed to regurgitate any of these fluids to the larvae (Gotwald 1974a).

Army ant workers do possess ovaries with maturing ova, but we do not know whether larvae consume eggs produced by workers as happens in other ant species (Gotwald 1971, Gotwald and Schaefer 1982). Schneirla (1971) speculated that when an *Eciton* queen produces a sexual brood, she also lays a batch of superfluous, nonviable eggs that are eaten by the sexual larvae.

When, how, and what do the adult workers eat? Wheeler and Bailey (1925) proposed that workers consume the liquids expressed from prey tissue when they form food pellets for the larvae. Many foraging driver ant workers return to the nest with their crops laden with prey fluids, and it is not unreasonable to assume that some portion of these fluids finds its way into their stomachs. Trophallactic exchange of food between adult workers (and for that matter, between workers and larvae) remains to be fully demonstrated as an important ingredient in the trophic and social life of army ants. Rettenmeyer (1963b) observed liquid food exchange between *E. dulcius* workers after they were fed wasp larvae and pupae. Both workers held open their mandibles while a droplet of fluid was cradled by the maxillo-labial complex (see Figure 3.3) of the donor worker. The exchange was followed by mutual cleaning behavior (Rettenmeyer 1963b).

It is clear that army ant workers sometimes consume their own brood, but the extent to which cannibalism occurs and the importance of the brood as a food source are not known. Certainly, brood cannibalism is common in ants, and the brood may serve as an emergency nutrient source in times of food shortage (Wilson 1971a). Rettenmeyer (1963b) compared the numbers of eggs and young larvae with the numbers of mature larvae and pupae in *E. hamatum* and discovered that a significant decrease occurred during the brood maturation period (he suggested that an egg brood range of 50,000 to 200,000 decreased to about 10,000 to 60,000 full-grown larvae). He concluded that cannibalism of worker brood may be extensive. There is little doubt that *Eciton* workers' cannibalism of sexual brood early in the brood's development helps keep the number of large, ravenous sexual larvae at a manageable level (Schneirla 1971).

Diet

For the most part, dietary observations of army ants are anecdotal, seldom quantified, and emphasize the sensational. With the exception of *Dorylus* (*Alaopone*) *orientalis*, a plant eater with a reputation for damaging agricultural crops (see Chapter 6), army ants are uncompromisingly carnivorous. To generalize further, there can be little doubt that the principal army ant prey are, in descending order of importance, ants, termites, and wasps (Table 4.1). This bill of fare is generously supplemented with a wide variety of other invertebrates and occasional vertebrates in those species that qualify as general predators.

Only a few army ant species are trophic generalists, versatile hunters of the forest floor. In the New World, most prominent among the rather indiscriminate predators are *Eciton burchelli*, *Labidus coecus*, and *L. praedator*; in the Old World, *Aenictus gracilis*, *A. laeviceps*, *Dorylus* (*Anomma*) *molestus*, *D.* (*A.*) *nigricans*, and *D.* (*A.*) *wilverthi* (Table 4.1). The vast majority are trophic specialists or oligophagous. An expanded diet, as exemplified in the polyphagous species, is correlated to some extent with epigaeic foraging and swarm raiding (Gotwald 1982). Conversely, trophic specialists tend to be hypogaeic column raiders.

For example, *Aenictus* species of both Africa and Asia, except for the epigaeic species *A. gracilis* and *A. laeviceps*, prey exclusively on the immature and adult stages of other ants (Brauns 1901, Crawley and Jacobson 1924, Sudd 1959, J. W. Chapman 1964, Gotwald 1976, Rosciszewski and Maschwitz 1994). Hypogaeic species of *Dorylus* are also specialists, primarily on termites or other ants. Although the nests of the social insects on which these specialists prey are scattered about the trophophoric arena, requiring a greater energy investment on the part of the searching predator, once found they constitute a wealth of concentrated food, well worth the energy expended looking for them (Carroll and

Table 4.1. Observed diets of selected New and Old World army ants

Species	Diet
	Aenictinae
Aenictus asantei	Adults and immatures of the ant genus *Pheidole* (Campione et al. 1983)
Aenictus binghami	Mainly brood and adults of ants (T. C. Schneirla, pers. comm., cited in Wilson 1964)
Aenictus ceylonicus	Observed attacking a colony of *Pheidole concinna* (R. H. Crozier, pers. comm., cited in Wilson 1964)
Aenictus eugenii	Specialized predator of ants, especially immature stages of the subfamily Myrmicinae (Gotwald and Cunningham–van Someren 1976); honeydew from *Pseudococcus* sp. (Santschi 1933)
Aenictus gracilis	Primarily ants (from 16 genera)—larvae, pupae, and adults; larvae, pupae, and callow adults of the social wasp *Ropalidia flavopicta*; myriapods; termites; small staphylinid beetles (J. W. Chapman 1964); a wide variety of invertebrates; large social wasps; adult insects; uncooked corn grits (Schneirla and Reyes 1966); ants, both brood and adults, of the genera *Technomyrmex, Paratrechina, Acropyga,* and *Prenolepis* (Rosciszewski and Maschwitz 1994)
Aenictus laeviceps	Primarily ants (from 16 genera)—larvae, pupae, and adults; larvae, pupae, and callow adults of the social wasp *Ropalidia flavopicta*; myriapods; termites; small staphylinid beetles (J. W. Chapman 1964); a wide variety of invertebrates; large social wasps; adult insects (Schneirla and Reyes 1966); ants, both brood and adults, of the genera *Pheidole, Polyrhachis, Camponotus, Crematogaster,* and *Prenolepis* (Rosciszewski and Maschwitz 1994)
	Ecitoninae
Eciton burchelli	All kinds of arthropods, mostly the brood of nonecitonine ants; polybiine and polistine wasps, both immature stages and adults; Orthoptera (sensu lato), including the families Blattidae (cockroaches), Gryllidae (crickets), and Tettigoniidae (long-horned grasshoppers); true bugs (Hemiptera); beetles (Coleoptera); rarely dragonflies (Odonata); a few termites (Isoptera); numerous spiders, especially Lycosidae; scorpions; a dead snake (Rettenmeyer 1963b); snakes, lizards, nestling birds (Schneirla 1956); social wasps (Chadab-Crepet and Rettenmeyer 1982); the social paper wasp *Polistes erythrocephalus* (Young 1979); social wasps (50 percent of the diet) (Rettenmeyer et al. 1983); the social wasp *Agelaia yepocapa* (O'Donnell and Jeanne 1990); crickets and cockroaches; ants, including the ponerine *Ectatomma ruidum* (Otis et al. 1986)

Table 4.1—*cont.*

Species	Diet
	Ecitoninae
Eciton dulcius	Mostly ants, especially of the subfamily Ponerinae (Rettenmeyer 1963b)
Eciton hamatum	Immature stages, occasionally adults, of wasps, primarily social Polybiinae and Polistinae (Vespidae); immature stages, occasionally adults, of noneciton- ine ants (Rettenmeyer 1963b); in Ecuador, almost exclusively ants, especially of the subfamilies For- micinae and Dolichoderinae; treehoppers (Homop- tera, Membracidae) tended by ants of the genus *Pheidole* (although these are carried to the nest, they may not be eaten); social wasps (Vespidae, Polybi- ini) (Rettenmeyer et al. 1983); social wasps (Chadab- Crepet and Rettenmeyer 1982)
Eciton mexicanum	Mostly larvae and pupae of ants, occasionally adults (Rettenmeyer 1963b)
Eciton rapax	Primarily large species of ants of the subfamily Poner- inae, including *Pachycondyla crassinoda* (Rettenmeyer et al. 1983); social wasps (Chadab-Crepet and Ret- tenmeyer 1982); more than 96 percent of prey items from the forest floor and understory are ants (larvae, pupae, adult workers, and other females and males), especially of the genera *Camponotus, Odontomachus,* and *Pachycondyla* (Burton and Franks 1985)
Eciton vagans	Primarily ants, immatures and adults (Rettenmeyer 1963b); *Polistes* wasps (Fiebrig 1907, cited in Retten- meyer 1963)
Labidus coecus	Eats a greater variety of items than any other Eciton- ini; ants (but not as extensively as other ecitonines); Amphipoda (beach fleas) (Crustacea); spiders; scar- ab beetles; Cicadidae (cicadas); Mantidae (praying mantids); Tettigoniidae; numerous moths belonging to a variety of families (Rettenmeyer 1963b); insect larvae in and under logs, under cow dung, and in the dead carcasses of cats and dogs (W. M. Wheeler 1910); fly larvae (Diptera), including the screw- worm, *Cochliomyia americana* (Lindquist 1942); seeds and nut meats (Borgmeier 1955); a Brazilian crab, *Trichodactylus argentinianus* (Lenko 1969)
Labidus praedator	A wide variety of arthropods, including Isopoda (sow- bugs) (Crustacea), Amphipoda, Araneida, Gryl- lidae (crickets), Tettigoniidae, Blattidae, and ants (Rettenmeyer 1963); a mouse (Reh 1897, cited in Rettenmeyer 1963b); carcasses of dead animals; sugar; crushed pineapple; boiled rice; palm fruit; dried apple (Borgmeier 1955); social wasps (Chadab- Crepet and Rettenmeyer 1982); larvae and pupae of

Table 4.1—*cont.*

Species	Diet
	Ecitoninae
	the leaf-cutter ant *Acromyrmex crassispinus* (Fowler 1977)
Neivamyrmex harrisi	Preys exclusively on the myrmecine ant *Solenopsis xyloni* (Mirenda et al. 1980)
Neivamyrmex nigrescens	Primarily ants, some beetles (Rettenmeyer 1963b); probably the ant *Novomessor albisetosus* (McDonald and Topoff 1986); the ant *Pheidole desertorum* (Droual 1984); preys exclusively on ants (mostly immature stages) and termites, with a clear preference for ants of the genus *Pheidole* (16 species of ants and 1 termite, *Gnathamitermes* sp.) (Mirenda et al. 1980)
Neivamyrmex opacithorax	Almost exclusively ants and ground beetles, family Carabidae (Rettenmeyer 1963b); small carabid beetles (W. M. Wheeler and Long 1901)
Neivamyrmex pilosus	Almost exclusively ants—adults, larvae, and pupae; a small spider (Rettenmeyer 1963b); ants, especially of the genus *Crematogaster* (Borgmeier 1955)
Nomamyrmex esenbecki	Ants, with 80 to 90 percent of retrieved prey consisting of immature stages (Rettenmeyer 1963b); ants, especially the ponerine *Odontomachus;* larvae, pupae, and callow workers of the leaf-cutter ant *Atta mexicana* (Rettenmeyer et al. 1983); social wasps (Chadab-Crepet and Rettenmeyer 1982)
	Dorylinae
Dorylus (Alaopone) orientalis	Roots and tubers of plants, including peanuts, potatoes, and turnips (see Chapter 6 for a complete inventory of plants eaten); earthworms; a beetle larva (grub) (Mukerji 1933)
Dorylus (Anomma) gerstaeckeri	Earthworms (Gotwald 1974a)
Dorylus (Anomma) molestus	Millipedes; ticks; insects of the orders Orthoptera, Hemiptera, Coleoptera, Lepidoptera, and Diptera (Swynnerton 1915); grasshoppers; crickets; pentatomid bugs (Hemiptera); a chameleon; geckos; a caged crocodile (Loveridge 1922); insects constitute 74 percent of diet: Lepidoptera, 53.7 percent; Coleoptera, 11.6 percent; Orthoptera, 8.7 percent; Hemiptera, 7.9 percent; other insects include Collembola (springtails), Diptera (flies), Homoptera, Hymenoptera, Isoptera (termites), and Neuroptera; a chameleon (*Chamaeleo jacksonii*); earthworms; arachnids; sowbugs; centipedes; millipedes (Gotwald 1974a)
Dorylus (Anomma) nigricans	Spiders; cockroaches, including their egg cases; grasshoppers; crickets; fly and ant pupae (Cohic 1948); termites (Bequaert 1913); insects constitute 67.2 percent of diet: Orthoptera, 40.3 percent; Hymenoptera,

Table 4.1—*cont.*

Species	Diet
	Dorylinae
	24.4 percent; Coleoptera, 12.2 percent; Lepidoptera, 6.7 percent; other insects include Dermaptera (earwigs), Diptera, Embioptera (webspinners), Hemiptera, Homoptera, Neuroptera, and Thysanura (silverfish); earthworms; centipedes; millipedes; arachnids; a caged snake; fallen palm nuts (*Elaeis quineensis*), from which they harvest the pithy outer covering; pieces of corn (maize) cob (Gotwald 1974a)
Dorylus (Anomma) wilverthi	Alata termites (Burgeon 1924a); insects predominate, including grasshoppers and other Orthoptera, mantids, cockroaches, Hemiptera, Homoptera, caterpillars and a chrysalis (Lepidoptera), beetles, hymenopterous pupae and adults, ants, flies (Diptera), earwigs (Dermaptera), and a scale insect; also spiders; sowbugs; plant seeds (Raignier and van Boven 1955)
Dorylus (Dorylus) helvolus	The maize stalk borer (Lepidoptera) (Moore 1913); the pouched mouse, *Saccostomus campestris* (Ellison 1988)
Dorylus (Typhlopone) juvenculus	The termites *Macrotermes michaelseni* and *Microcerotermes* sp. (Darlington 1985); honeydew from membracid (Homoptera) nymphs that were feeding on roots of maize (Arnold 1915)
Dorylus (Typhlopone) dentifrons	The termite *Macrotermes bellicosus* (Bodot 1961, 1967); the termite *Acanthotermes spiniger* (Hegh, cited in W. M. Wheeler 1936)
Dorylus (Typhlopone) labiatus	The termite *Microtermes mycophagus* (Sharma and Bohra 1968)

Janzen 1973). *E. hamatum*, which preys on ants and wasps (see Figure 6.4), is epigaeic and therefore does not conform to the hypogaeic-specialist correlation. But this species is a column raider and does occasionally take other insect prey (Rettenmeyer 1963b). Other specialists such as *Nomamyrmex esenbecki* and *Neivamyrmex pilosus* are column raiders and negatively phototaxic, although they will forage on the soil surface.

Certain of the African driver ants—for example, *Dorylus (Anomma) molestus* and *D. (A.) nigricans*—are unquestionably the most catholic of the generalists, even though they nest in the soil. In the New World, only *E. burchelli*, *L. coecus*, and *L. praedator* approach the level of polyphagy achieved by the driver ants. These ecitonine species are swarm raiders—though Rettenmeyer (1963b) described *L. coecus's* foraging pattern as intermediate between column and swarm raiding—and all

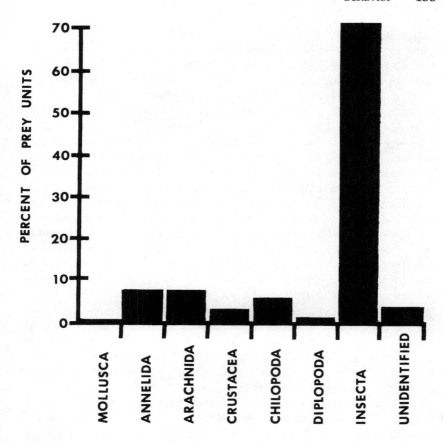

Figure 4.9. Taxonomic distribution of prey units retrieved by three species of driver ants (*Dorylus*) in Ghana and Kenya. Although insects dominate in the composite data, one of the driver ant species, *D.* (*Anomma*) *gerstaeckeri*, feeds exclusively on earthworms (see Figure 4.10). (Reprinted, by permission of the Entomological Society of America, from Gotwald 1974a.)

raid on the substrate surface, at least occasionally. Although *A. gracilis* and *A. laeviceps* are epigaeic column raiders, they are trophic generalists with a predilection for other ants (Schneirla and Reyes 1966).

What can we conclude from these observations on foraging behavior and diet? It would seem that army ants that have retained the ancestral life mode are hypogaeic nesters, hypogaeic column raiders, and specialist predators on ants. Army ants have a demonstrated tendency to become surface foragers, and in doing so to expand their diet. If the epigaeic forager retains its column raiding pattern, its dietary expansion is limited. If, on the other hand, the epigaeic forager becomes a swarm raider, its diet expands to include a wide variety of invertebrates and some vertebrates. It is plausible that swarm raiding and polyphagous feeding make possible extraordinarily large colonies, as the driver ants seem to demonstrate (Gotwald 1982). The polyphagous predator not only capitalizes on a diverse prey menu available on a daily basis but can also exploit food sources that are abundant only periodically and serendipitously, such as emerging alate termites.

How food preferences of the polyphagous army ants relate to their habitat and to prey availability, life stage, and size has been largely ignored by field researchers. A comprehensive investigation of this kind exists only for three species of driver ants: *D. (A.) gerstaeckeri, D. (A.) molestus*, and *D. (A.) nigricans* (Gotwald 1974a). Although these ants prey on a wide variety of invertebrates, insects form the largest portion of their diet (Figure 4.9). Earthworms, especially, and arachnids (primarily spiders and scorpions) follow in importance (Gotwald 1974a). The percentage of insects represented in driver ant diets may merely reflect the relative abundance of insects on the forest and savanna floor, although *gerstaeckeri* seems to feed solely on earthworms (Figure 4.10). It may indeed be a narrow specialist. Thirteen insect orders were represented among the prey collected by *molestus* and *nigricans* (Figure 4.11). Although Lepidoptera were most commonly taken by *molestus* and Orthoptera by *nigricans*, the most noteworthy difference between the two was their relative exploitation of the Hymenoptera, which were rare among the prey units retrieved by *molestus.*

Analysis of prey preferences of *molestus* and *nigricans* according to habitat in Ghana and Kenya revealed that relative prey abundance was a function of habitat type and was reflected in the prey collected (Figures 4.12, 4.13). Insect prey units were also classified according to the developmental stage they represented. Of the 945 prey units collected from the foraging workers, 590 (or 61.8 percent) were from immature stages. A majority of the prey units were from holometabolous insects. It thus seemed clear that prey vulnerability must strongly affect the composition of the diet (Gotwald 1974a). Caterpillars, ant larvae, and holometabolous insect pupae are presumably more likely to be captured than are fleet-footed or winged adults.

My 1982 analysis of prey preferences also considered prey unit size. The term *prey unit* implies that large prey organisms are sectioned into fragments by foraging ants before being transported back to the nest. Even small prey that are not sectioned may be trimmed of legs and antennae. Prey units of slugs, earthworms, arachnids, millipedes, and insects were confined to a range of 1 to 15 mm. Centipede units were in the range of 20 to 30 mm. Thus, two distinct nonoverlapping size

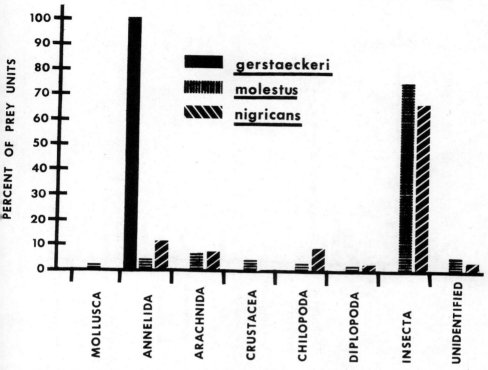

Figure 4.10. Comparative diets of three species of driver ants (*Dorylus*) in Ghana and Kenya. (Reprinted, by permission of the Entomological Society of America, from Gotwald 1974a.)

ranges for prey units existed. Although centipede prey units were also observed in the smaller size range, they sometimes constituted the longest prey units of all (the whole centipede) and were transported by two or more workers in tandem (Gotwald 1982).

Two reports of army ants gathering honeydew from homopterous insects have been published. *Dorylus* (*Typhlopone*) *fulvus* was observed tending immature membracids (treehoppers) on the roots of maize (Arnold 1915), and the East African species *Aenictus eugenii* was once collected while tending a species of *Pseudococcus* (Santschi 1933). Honeydew, a mixture of sugars, amino acids, amides, proteins, minerals, and vitamins, is the modified phloem sap of plants that has passed through the gut of sap-feeding Homoptera (aphids, mealybugs, treehoppers, etc.). Droplets of this fluid, which appear at the insect's anus, are solicited and collected almost exclusively by ants of the subfamilies Myrmicinae,

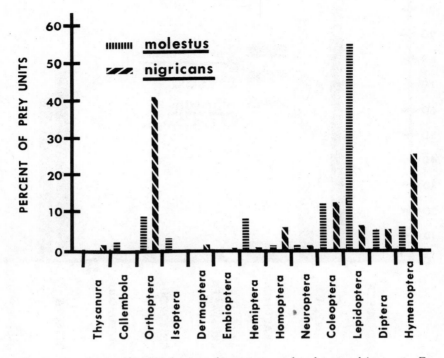

Figure 4.11. Taxonomic distribution of insect prey taken by two driver ants, *Dorylus* (*Anomma*) *molestus* and *D.* (*A.*) *nigricans*. Thirteen orders of insects are represented. (Reprinted, by permission of the Entomological Society of America, from Gotwald 1974a.)

Dolichoderinae, and Formicinae (Hölldobler and Wilson 1990). That army ants do not commonly collect honeydew could be anticipated; that they do it at all is remarkable.

Emigration

Nest relocation is more common among ants in general than myrmecologists once thought. In some ants—including army ants (which may deplete local food resources), fugitive species (those that nest in fragile, quickly deteriorating sites), and tramp species (for whom quick saturation of disturbed areas is adaptive)—emigratory behavior evolved reciprocally with other specializations (Smallwood 1982). But emigration and nest relocation also occur in species whose habits do not require colony movement, at least in any obvious way; *Formica*, *Aphaenogaster*, and *Tapinoma* species in the deciduous forests of the eastern United

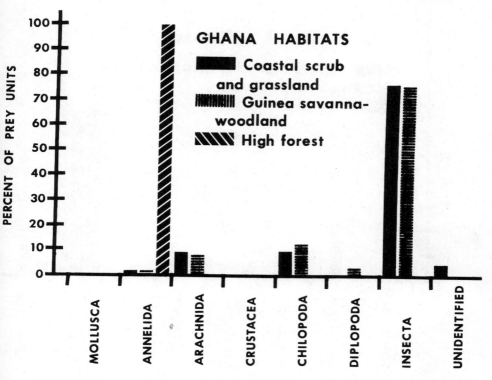

Figure 4.12. Taxonomic distribution of invertebrate prey, arranged according to habitat, taken by two species of driver ants, *Dorylus* (*Anomma*) *gerstaeckeri* and *D.* (*A.*) *nigricans*, in Ghana. Data from the coastal scrub and grassland and Guinea savanna-woodland represent prey units taken only by *nigricans*; in the high forest, the major prey item was the earthworm, for both *gerstaeckeri*, an earthworm specialist, and *nigricans*, a polyphagous species, revealing that the relative abundance of prey may be a function of habitat type. (Reprinted, by permission of the Entomological Society of America, from Gotwald 1974a.)

States are examples. By emigrating, ant colonies may avoid overshading by growing vegetation, decreases in soil moisture, nest fungi, and parasites; or by moving they may elude discovery by predators or reduce competition with other ant colonies (Smallwood 1982).

Undoubtedly, emigration behavior in army ants evolved with, and as a consequence of, group predation. In nonphasic army ant species, especially general predators, however, regular nest movement may not even occur. Only phasic species can be regarded as predictable in their emigrations to new nest sites. The nomadic phase of their functional reproductive cycle begins with the eclosion of callow workers and is driven by the insatiable hunger of the rapidly growing larvae. Topoff

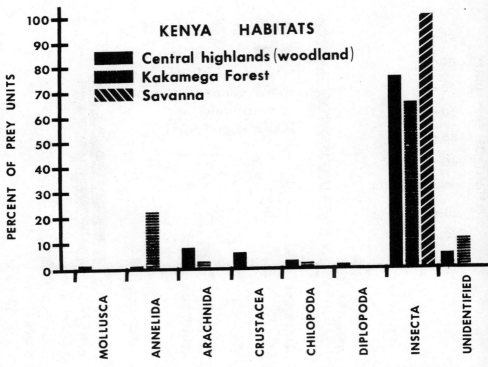

Figure 4.13. Taxonomic distribution of invertebrate prey, arranged according to habitat, taken by the driver ant *Dorylus* (*Anomma*) *molestus* in Kenya. Although insects constitute the major prey taken in all three habitats, earthworms are important dietary items in the Kakamega Forest habitat, where they are abundant. (Reprinted, by permission of the Entomological Society of America, from Gotwald 1974a.)

and Mirenda (1980) demonstrated empirically that larval hunger in *Neivamyrmex nigrescens* may prod the colony to its daily emigrations once the nomadic phase actually begins. Overfed laboratory colonies, for example, emigrated less than colonies given little food. The primacy of food scarcity among the proximate causes of the nomadic phase is rather clear. Most likely, raids during the nomadic phase do not bring in enough prey to satiate the larvae.

Although emigration was the focus of much of Schneirla's exhaustive research (e.g., 1938, 1944, 1945), other early observers characterized the fascinating vagabond nature of army ant existence as well. Describing sorties of *Eciton* in Mexico, F. Sumichrast (1868, p. 40) said that "one can believe them to be sometimes expeditions of pillage, sometimes changes of domicile, veritable migrations." In Nicaragua, Thomas Belt (1874, p.

23) observed that "the Ecitons are singular amongst the ants in this respect, that they have no fixed habitations, but move on from one place to another, as they exhaust the hunting grounds around them. I think *Eciton hamata* does not stay more than four or five days in one place. I have sometimes come across the migratory columns. They may easily be known by all the common workers moving in one direction, many of them carrying the larvae and pupae carefully in their jaws." In Africa, Savage (1847, p. 4) commented on driver ants "appearing and disappearing from certain localities." So did J. Vosseler (1905), who speculated that colony emigrations were stimulated by the shortage of food.

Eciton colonies always conduct day-long foraging expeditions before emigrating (Schneirla 1938, 1945, 1971). As *E. burchelli* and *E. hamatum* raiding progresses during the afternoon, three conditions presage an impending emigration: a high level of excitement in the colony that guarantees a continued exodus of foraging workers from the bivouac, traffic complications among workers on the raiding trails that prevent a general return of foragers to the bivouac, and environmental changes at dusk that trigger a behavioral shift in the raiding workers from foraging activities to actions essential to emigrations (Schneirla 1971). In *E. hamatum*, one of the three main raiding trails serves as the emigration route, and this trail teems with workers leaving the bivouac and prey-laden foragers returning from raids. The returning workers are intercepted and sucked into the outward-moving stream (Schneirla 1938). The exodus becomes an emigration when workers begin carrying brood from the bivouac.

While some workers are still leaving the old bivouac, others are forming the new one, a process that often begins between 6:00 and 8:00 P.M. in *E. hamatum* and between 7:30 and 9:30 P.M. in *E. burchelli* (Schneirla 1971). When transport of the brood is more than half completed, the queen emerges with a retinue of attentive workers (Figure 4.3). In the early nomadic phase, this occurs between 7:00 and 8:00 P.M. in *E. hamatum* and between 8:00 and 10:00 P.M. in *E. burchelli* (Schneirla 1971). Rettenmeyer (1963b) found that the size of the *E. hamatum* queen's retinue depends primarily on the speed with which the queen travels. If her progress is unimpeded, the retinue is small, consisting of about 50 workers and majors (Rettenmeyer et al. 1978).

The largest retinue Rettenmeyer (1963b) observed included between 25 and 50 major workers and a larger number of smaller workers, which stayed within 15 to 30 cm of the queen. The queen's retinue in *Eciton* and *Neivamyrmex* is not a fixed group of individuals that remains with her throughout an emigration. Membership in the retinue changes constantly. Nevertheless, major workers constitute a greater percentage of the queen's retinue than they represent in the colony. In *Aenictus laeviceps*, the queen's entourage is usually 5 to 8 cm at its widest point and as long as 1 meter (Schneirla and Reyes 1969). Probably, the retinue protects the queen against predators and various environmental hazards and so is largest in army ant species that nest and forage epigaeically (Rettenmeyer et al. 1978).

The general pattern in *Neivamyrmex nigrescens* is similar to that of *E. hamatum,* except that the former may forgo emigration on some nights during the nomadic phase (Mirenda and Topoff 1980). Schneirla (1958) observed 60 *N. nigrescens* emigrations. All the moves occurred at night, and all grew out of raiding activities. Emigrations usually began before midnight, about six hours after the beginning of raiding (which is also crepuscular and nocturnal). Each emigration required about six to eight hours to complete, so that most were over at dawn. In some instances, when interrupted by heavy rain, for example, an emigration required a second night to complete. Schneirla (1958) found that emigrations always occurred over major foraging trails. The distances covered ranged from 2.5 to 76 meters. The *N. nigrescens* queen appears to leave the nest earlier than does the *E. hamatum* queen; she enters the column when one-third of the emigration is completed (Schneirla 1958).

Predictably, less is known about the emigrations of more cryptic, subterranean species of New World army ants. *Labidus praedator,* for example, was observed emigrating by Schneirla (1971) only four times. The emigration columns are commonly 4 to 10 ants wide and include thousands of callow workers. Large numbers of worker cocoons are carried. This species constructs walls of earth pellets that flank the columns and sometimes form arcades completely shielding emigrating workers (Schneirla 1957; Rettenmeyer 1963b). Rettenmeyer (1963b) observed an emigration of this species that required more than one day to complete. Only a short portion of the emigration column (about 4 meters) moved on the soil surface, and even then it was shielded by soil particle walls and arcades and by workers themselves. *Cheliomyrmex megalonyx,* a distinctly subterranean species, was also observed to build soil particle arcades when emigrating across open spaces (W. M. Wheeler 1921).

Emigrations in *Aenictus gracilis* and *A. laeviceps*—both phasic epigaeic species—are usually initiated along major raiding columns that have been in progress for a considerable time. Emigrations in these two species can begin within 20 minutes of the first signs of excitement in the bivouac and may occur without the raiding precondition evident in *Eciton.* Furthermore, emigrations may begin at any time of day or night, early or late in raiding, as actions overlapping previous emigrations or as events ending inactive intervals and without the precondition of extranidal activity (Schneirla and Reyes 1969). The fact that emigration in these two species can arise either from raiding or from other colony activities prompted

Schneirla and Reyes (1969) to propose that the *Aenictus* pattern may represent a generalized, primitive condition in colony organization.

During the first emigrations of the nomadic phase in *A. gracilis* and *A. laeviceps*, the larvae, which are quite small, are carried in packets by the workers; consequently, the entire brood may be transported out of the old bivouac in less than 20 minutes. The larvae, not the emergence of callow workers, appear to precipitate the nomadic phase in these species. Later in the nomadic phase, when the larvae are larger, each is carried by an individual worker, increasing the time required to remove the larvae from the bivouac. Generally, the queen leaves the old bivouac during the second half of the emigration and is accompanied by a retinue of workers (J. W. Chapman 1964, Schneirla and Reyes 1969). Most unusually, both species may undertake more than one emigration per day. Early in the nomadic phase the emigrations last from two to three hours, later they may take four to seven hours (Schneirla and Reyes 1969).

Emigrations of African *Aenictus* have not been described in detail. Most significantly, their frequency and the nature of the functional reproductive cycle remain unknown. Presumably, most *Aenictus* species are nonphasic, although the lack of pupae in emigration columns hints of some correlation between brood development and emigration episodes. I observed an emigration of *A. asantei* at 9:50 A.M. crossing a path between plots of cassava. Except for an occasional worker, the column was unidirectional and approximately six ants wide. At 10:10, the queen, whose gaster was contracted, passed with a small entourage of workers (Gotwald 1976).

Of the extensively studied army ants, the African driver ants are the most erratic in their emigrations. Raignier and van Boven (1955) observed that driver ant emigrations often follow previously used foraging trails and that emigration columns are more often subterranean than are foraging columns. An emigration is a single episode (mean distance covered, 223 meters) that may take 2 or 3 days (or more) to complete (mean duration, 56 hours) and may not be followed by another emigration for as long as 125 days (this was an extreme case recorded for D. [A.] *nigricans* [Raignier and van Boven 1955]). Leroux (1982) calculated the average nest occupation at 8 days for *D. (A.) nigricans* and found no fixed pattern of emigrations. One colony, for instance, conducted 51 emigrations in one year. Similarly, a colony of *D. (A.) molestus* was found to have an average nest stay of 13.4 days, with a minimum stay of 3 days and a maximum stay of 45 days (Gotwald and Cunningham–van Someren, 1990) (Figure 4.14).

Raignier and van Boven (1955) noted that emigration columns can move along more rapidly than foraging columns and calculated that some moved as fast as 155 meters in five hours. Gotwald and Cunning-ham–van Someren (1990) discovered that some *D. (A.) molestus* colonies emigrate only short distances and reuse emigration trails and nest sites (Figure 4.14). In fact, different conspecific colonies may utilize the same trails and nests at different times. For example, one of those nests was occupied on fifteen separate occasions by four different colonies. And Leroux (1982) discovered that colonies of *D. (A.) nigricans* may emigrate at the same time that they conduct raids but are more likely to emigrate following raiding (74 percent of the time). This species also may emigrate without a precipitating raid (Plate 16).

As noted previously, statary *Eciton burchelli* colonies rotate successive raids around the central nest site in a nonrandom fashion (Franks and Fletcher 1983). Nomadic navigation in this species is also nonrandom. Franks and Fletcher (1983) found that the angles between successive emigrations of *E. burchelli* colonies on Barro Colorado Island had a mean of 6.22° and a standard deviation of 51.73°. The mean was not signifi-cantly different from 0°; therefore each nomadic raid and subsequent emigration was confined to approximately the same compass bearing as the previous day's raid and emigration (Franks and Fletcher 1983). This navigational behavior results in greater separation of statary bivouacs than would be achieved if colonies raided and emigrated in random directions. Ants that move nonrandomly may avoid raiding previously searched areas, and nomadic phase colonies can also avoid foraging are-nas swept of prey during former statary phases. Only when a colony fails to emigrate on one day does it abandon the direction established on the preceding emigrations. Franks and Fletcher (1983) attributed non-random navigation behavior to selection for colonies that avoid colli-sions with conspecific colonies.

Nomadic phase *Neivamyrmex nigrescens,* on the other hand, conduct emigrations that are unpredictable in both direction and distance (Mir-enda and Topoff 1980). Emigration direction and distance in this species are likely affected more by distribution and availability of prey and nest sites than by brood stimulation (Mirenda and Topoff 1980). It is not unusual for nomadic phase colonies of *N. nigrescens* to stay in one area

Figure 4.14. Emigratory history of a colony (KC-506) of *Dorylus* (*Anomma*) *molestus* observed for 432 consecutive days. Movements were limited to approximately 5 hectares of diverse but highly modified habitat (A). The colony emigrated 38 times (B) and revisited some nest sites several times (especially 075B). Nest sites are represented by dots, and emigrations are numbered consecutively next to arrows indicating the emigration routes. (See Gotwald and Cunningham–van Someren 1990.)

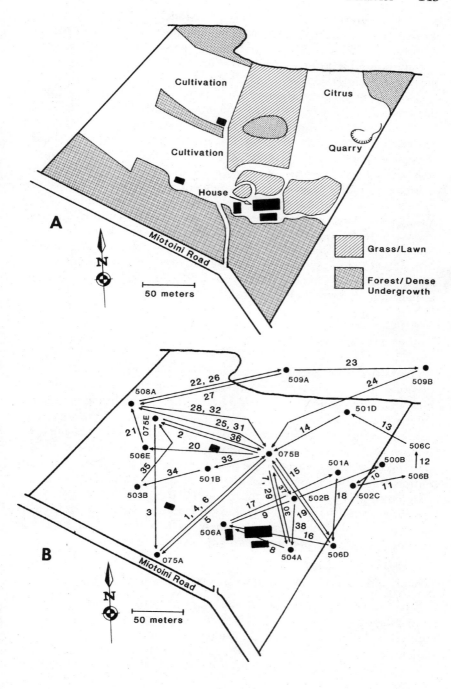

for several days, to return to earlier exploited areas, or to find themselves within a few meters of another nomadic phase colony. Mirenda and Topoff did find a positive correlation between the number of larvae and the distance emigrated.

G. R. Cunningham–van Someren and I (1990) were also unable to demonstrate any form of regular navigation in *Dorylus* (*Anomma*) *molestus*. Although some emigrations did allow access to foraging areas not recently exploited, often they resulted in substantial overlap of successive trophophoric fields. Colony KC-506 conducted thirty-one foraging raids from one nest before emigrating to another. Of the fifteen raids mounted from this second nest, 26.7 percent overlapped with raids conducted from the previous nest. Two subsequent emigrations produced foraging overlaps of 30 percent and 25 percent. The same colony conducted consecutive raids that covered the same area, and when it coexisted with other conspecific colonies in the research area, it commonly overlapped the foraging areas of those colonies. This driver ant species is a more general predator than *E. burchelli*, which perhaps explains its rather random navigation patterns.

Colony Fission

Unlike the majority of ant species, in which colonies are founded by individual queens, new army ant colonies are founded when existing colonies divide into daughter colonies. This form of colony founding—called colony fission, budding, or swarming—is also found in honey bees and many polybiine wasps (Bulmer 1983). Because the army ant queen is wingless, she makes no nuptial flight, and insemination occurs under rather obscure circumstances (see Mating, below). Essentially, half the dividing army ant colony follows the parent queen (unless she is superceded by one of her daughter queens) and half goes with a new daughter queen, a process that produces two ready-made colonies. Colony fission is clearly an adaptative correlate of group predation, since successful group raiding and retrieval of prey require colonies of a critical size.

Colony Fission and Kin Selection Theory

Colony fission constitutes an interesting theoretical challenge to our understanding of kin selection, the prevailing hypothesis explaining the presence of sterile castes in the social Hymenoptera. Kin selection theory proposes that individuals may altruistically reduce their personal survival and reproduction on behalf of other individuals with whom they share genes by reason of common descent, because in the process of enhancing their relatives' survival they promote the perpetuation of their own genes (Hamilton 1972). Colony fission poses a problem to kin selection theory: Possible differences between the inclusive fitness of workers that stay with the parent queen and those that go with the daughter queen suggest a genetic basis for worker-worker conflict over who will stay with the parent queen and who will go with the daughter queen (Macevicz 1979).

There is some consensus among biologists, emanating from the efforts of R. L. Trivers and H. Hare (1976) to predict the ratios of investment between male and female reproductives in the social Hymenoptera, that different members of a colony may have different preferred sexual investment ratios. For instance, it is predicted that workers will favor investment of resources in new queens over investment in males (by a ratio of 3:1) because they have a greater number of genes in common with the new queens, their sisters, than with new males, their brothers (with whom they share but one-fourth of their genes). Thus, according to R. A. Fisher, the queen will favor a 1:1 sexual investment ratio as she shares half of her genes with both workers (her daughters) and males (her sons).

Inclusive fitness is the total measure of an individual's reproductive success plus its influence on the reproductive success of relatives—with whom, of course, it shares some genes (Hamilton 1972). Army ant workers that stay with their parent queen continue to care for their sisters, with whom they share, on average (a measure called the coefficient of relationship), three-fourths of their genes. Those that go with their new queen sister are consigned to raising their nieces, with whom they share only three-eighths of their genes. Thus, these latter workers may significantly reduce their inclusive fitness (Macevicz 1979).

Colony fission does not fit accepted theory in another way as well: In any reproductive brood of army ants, many more males are produced than females. In other words, investment in reproductives is strongly biased toward males. This contradicts Fisher's sex ratio principle (Fisher 1958) for haplodiploid organisms—those in which females are diploid but males arise from unfertilized eggs and are therefore haploid—which predicts an equal investment in male and female reproductives (Macevicz 1979, Bulmer 1983). Although neither of these concerns—the reduced inclusive fitness of some workers and the biased investment in the production of males—has been fully reconciled with altruism and kin selection theory, some suggested explanations are worth exploring.

Addressing the issue of the decreased inclusive fitness required of workers that choose to go with their queen sister during colony division, Stephen Macevicz (1979, p. 364) recognized two enigmas. First, how did such a system of reproduction evolve given the possible decrease in inclusive fitness exacted from some of the participants? Second, how do workers, if they control the means of colony reproduction (through their contributions to colony growth and the size of the entourage that accompanies each queen), surmount the apparent dilemma "that the relative increase in entourage size dictated by natural selection means that a greater number of workers will suffer in terms of inclusive fitness"?

If colony fission involves an altruistic act—the loss of inclusive fitness on the part of some workers—then fission can be explained in terms of either parental manipulation or kin selection (Macevicz 1979). Parental manipulation implies some form of control by the queen over worker choices or fates. Queen ants, for instance, may be able to make their offspring into workers instead of queens through starvation or hormone-mediated sterilization in situations in which relatedness favors a worker class (Hölldobler and Wilson 1990). If parental manipulation is responsible for worker altruism, then a 1:1 sexual investment ratio (under random mating) would be predicted, and questions concerning fitness are largely irrelevant (Macevicz 1979). The alternative explanation, kin selection, is very much about the inclusive fitness of the workers and who shall incur the loss by joining a queen sister. That issue remains unresolved.

Workers that do choose their queen sister over the maternal queen are eventually replaced, as they die, by their nieces. As the nieces, the daughters of the new queen, supplant the older generation, the average intracolony relatedness decreases, reaching the lowest point when the proportion of the two generations in the colony is 1:1. The average of relatedness increases from that moment on, as older workers continue to die, and will eventually approach the level that existed before the original colony divided (see Tobin, pers. comm., cited in Hölldobler and Wilson 1990, p. 186).

W. D. Hamilton (1975, p. 177), who developed and eloquently promoted the theory of kin selection in the 1960s and 1970s, attempted to explain the abundance of males and the paucity of females in army ant sexual broods as follows: "The 'female' unit of reproduction is the swarm itself—queen plus the workers necessary to make a viable unit—so if by Fisher's Principle we make the mass of males equal to the mass of emitted swarms we see that males are expected to be much more abundant than queens in genera like *Apis* and *Eciton*." In an analysis of sex ratio theory in social insects that reproduce by colony fission, however, M. G. Bulmer (1983) concluded that there is little evidence to support Hamilton's notion that the investment in the worker swarm can be counted in the investment in female reproductives. Instead, Bulmer favored Robin Craig's (1980) argument that since only one new female is permitted to survive colony fission, there is nothing to be gained by increasing the number of queens produced. The best strategy is for queens and workers to produce as many males as possible, because males may have an opportunity to mate with foreign queens and spread the colony's genes in that way.

Franks (1985) pointed out that the fitness of all members of an army ant colony depends on the colony's growth rate, which also determines the rate at which colonies can divide. Consequently, he noted, selection pressure should increase the rate of successive colony divisions by reducing colony generation times. That is, colonies should grow as quickly as possible and divide at a size that maximizes the growth rates of their daughter colonies. Furthermore, the parent colony ought to divide the worker population equally in two, as this will minimize the average generation time of the daughter colonies. The timing of colony fission will be a function of the growth rates of both parent and daughter colonies. Franks (1985) predicted that the combined growth rate of the daughter colonies should equal the growth rate of the parent colony in order for the parent colony to maximize its fitness, assuming that this colony divides in two equally. *Eciton burchelli,* according to Franks, appears to fit this prediction.

When the colony divides, workers choose either to stay with their mother or to go with a queen sister. In theory, workers should selfishly select the queen that will maximize the workers' inclusive fitness (Franks and Hölldobler 1987). But the choice is not as straightforward as it might seem. Workers that choose their sister queens resort to raising nieces and thereby reduce their inclusive fitness (as reviewed above). Eventually, the workers who stay behind will have to reject their maternal queen as a consequence of her senility. Finally, because army ant queens probably mate more than once, virgin queens may not be full sisters of the workers.

If they and the virgin queens are full sisters, workers should then choose the most fertile queen. If their maternal queen has mated more than once, however, then each patrilinial group of workers in the colony should prefer its own full sister queens to any stepsister queens. In order to choose correctly, the workers would have to be able to discriminate the degree of kinship they share with each queen—a skill not yet unequivocally demonstrated to exist in the ants. But the need for such a kin recognition skill in the army ants may actually be minimal. Because queens probably mate but once a year and because workers have a high mortality rate, it is quite likely that the new queen and the majority of workers are full sisters. Regardless of whether workers can recognize the degree of their relatedness to the queens, the ability to differentiate between queens on the basis of their potential fertility and survivorship will be the character selected for (Franks and Hölldobler 1987).

Queen choice by workers involves a variety of proximate cues and may begin in the colony's nursery. In *Eciton hamatum*, for instance, some mature queen larvae are more attractive to workers than others, and some advantage may accrue to these queens in the selective process that precedes colony division (Schneirla 1956). The nature of this attractiveness is unknown, although it is possible that the sexual larvae produce a brood pheromone (Franks and Hölldobler 1987). Certainly, once the new queens have eclosed, they produce true queen pheromones, which may also play a role in the selection process.

The Course of Colony Fission

The rearing of sexual brood is prerequisite to colony fission. The proximate stimulus for sexual brood production remains elusive. Schneirla (1971) maintained that the onset of dry weather in areas with distinct seasonal changes may be the stimulus, at least in the *Eciton, Neivamyrmex,* and *Aenictus* species he studied. Although Schneirla (1971) convincingly argued this hypothesis, Rettenmeyer (1963b) observed an absence of synchrony in the production of sexuals in *E. hamatum* colonies that inhabited the same area during the dry season. Indeed, some species of *Eciton* actually produced sexual broods during the rainy season (Rettenmeyer 1963b). Nevertheless, it was Schneirla's (1971) contention that the abrupt environmental change represented by the onset of the dry season radically affects the reproductive physiology of the queen. The physiological changes thus wrought, he reasoned, can inhibit fertilization (which must be suppressed for the production of males) and establish the pattern for sexual broods.

Schneirla (1971) postulated, in the absence of hard evidence, that the first group of eggs produced in a sexual brood is deposited just before the inhibition of fertilization. These fertilized eggs are destined to become, as a result of differential treatment and feeding, new queens. A second group of eggs is deposited unfertilized, giving rise to males. Schneirla (1971) further proposed a third group of eggs with a trophic function; that is, they serve as a specialized food source for the sexual larvae.

S. E. Flanders (1976) elaborated on Schneirla's speculations. He suggested that an enlargement in the duct that passes from the ovoid spermatheca of the *Eciton* queen (see Chapter 3) may be a reservoir for a sperm-activating secretion through which sperm cells must pass before being deposited on the eggs. This secretion, he proposed, clears the partly coiled spermathecal duct of residual spermatozoa that were trapped in the coils of the duct at the end of the previous egg-laying bout. It is these leftover sperm cells that fertilize the first eggs deposited in a sexual brood. Further sperm cells are not forthcoming from the spermatheca as a physiological consequence of the queen's abrupt exposure to the dry season. Thus, eggs deposited subsequent to the initial batch are not fertilized.

Schneirla (1956) divided colony division in *Eciton hamatum* and *E. burchelli,* into four stages. In the first stage, latent fission (p. 52), the brood, still in early stages of development, is reduced in numbers as workers devour up to 90 percent of the eggs. This results in persistent overfeeding of queen and male larvae, which represent about 8 percent of an all-worker brood. As a consequence, sexual larvae experience accelerated growth and a shorter larval stage than worker larvae.

During the early phases of larval development, sexual brood are highly attractive to their attendant workers. This appeal intensifies as development proceeds. Eventually, the placement of sexual brood in the bivouac is asymmetrical. In *E. hamatum,* for instance, the queen and brood are polarized, with the queen on one side of the bivouac and the brood concentrated on the opposite side. When an all-worker brood is present, on the other hand, immatures are distributed concentrically and symmetrically at the periphery of the bivouac, and the queen is in the upper center of the nest.

With polarization of the bivouac, a significant portion of the worker population becomes selectively affiliated with the sexual brood, while the remaining workers remain fixed in their attachment to the queen. In other words, the internal dynamics of the bivouac before fission involves development of differential allegiances: workers choose either to stay with the functional queen or to go with one of their new sister queens.

Curiously, if the functional queen is removed from the colony at this time and later reintroduced to her daughters, she may be treated ambivalently or even with hostility. In contrast, when a queen is removed from a bivouac with an all-worker brood present and then reintroduced, she is unconditionally accepted by any workers to whom she is presented. Workers in the dividing colony are distinguished by their uncertain but resolving fealties. Thus, the basic character of the prefission colony is bipolar, with the functional queen and the sexual brood acting as counterattractants (Schneirla 1956, p. 54).

Secondary allegiances develop among workers affiliated with the sexual brood. As the queen larvae near maturity, it is clear that some of the potential queens are more alluring to the workers than others. Those that are most attractive may ultimately be more likely to be selected as the new virgin queen at time of actual fission.

Schneirla's (1956, p. 54) second stage of colony fission occurs when the young queens emerge. The new queens, which invariably eclose from their cocoons before the males, never number more than six in both *E. hamatum* and *E. burchelli*. The later-emerging males, on the other hand, average about 1500 in the former species and 3000 in the latter. Because they emerge about three days earlier than the males, young *E. hamatum* queens may be overfed and may experience accelerated maturation as a result. At this time as well the workers become differentially habituated to the various callow females (Schneirla 1956). The first one or two queens to eclose enjoy the greatest probability of surviving to become queen in a daughter colony. Most likely these leading virgin queens were more attractive to workers as maturing larvae as well. Certainly the last new queen to emerge is predictably the least attractive to workers and not a realistic candidate to be a functional queen.

In *E. hamatum*, each virgin queen acquires an entourage of workers, numbering in the hundreds or low thousands, as she emerges from her cocoon. Within hours of emergence, the first of the young queens to eclose moves with her entourage to a position within 1 meter of the bivouac. The second virgin queen to emerge also moves out of the bivouac, but her entourage is smaller and she settles closer to the bivouac (Figure 4.15). These queens and their clustered workers remain close to the bivouac, although they may move a few centimeters farther away. Bidirectional columns of workers connect the queens and their affiliated workers to the nest, and prey items are carried from the bivouac to the clusters.

Later-emerging queens have a disturbing effect on workers, who confine them to the bivouac wall. This worker behavior Schneirla (1956, p. 57) termed the sealing-off reaction. From the moment of oviposition, potential queens run an ever-narrowing gauntlet of selection, until, in the end, one, or sometimes two, virgin queens survive to be crowned.

Before this, however, a third stage occurs: the eclosion of the males (Schneirla 1956, p. 58). Like the emergence of an all-worker brood, which energizes the statary colony of *Eciton* into a major raid, emigration, and a new nomadic phase, the eclosion of males anticipates the end of the statary phase and foretells new levels of worker activity. The stimulatory effect of emerging males is even greater than that of emerging workers and results in a multiple raiding system, from which devolve two or more major columns. These divergent foraging routes are essential to colony fission.

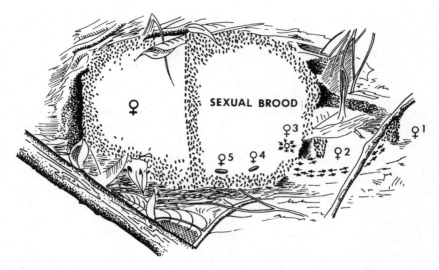

Figure 4.15. In the second stage of colony fission in *Eciton hamatum*, the new queens eclose from their cocoons; each attracts an entourage of workers; one or two of the queens, with their workers, leave the bivouac to form small clusters within 1 meter of the bivouac (1 and 2). Each cluster is connected to the bivouac by a two-way column of workers. A third (3), more recently emerged, queen is restrained by a group of workers in the bivouac wall. Other queens (4 and 5), still confined to their cocoons, are held by workers near the base of the bivouac. (Redrawn, by permission of Birxhäuser Verlag, from Schneirla 1956.)

Male eclosion is followed by the fourth and final stage, the overt division (Schneirla 1956, p. 59). The dividing colony establishes its foraging trails at daybreak. Usually, three major columns diverge from the bivouac. By day's end, the colony has split into two daughter colonies, each of which follows a different foraging trail. Most commonly, the old queen becomes the functional queen of one of the daughter colonies, and a virgin queen is the functional queen of the other. Sometimes the mother queen is superseded by one of the virgin queens, but Schneirla (1956) considered this a secondary form of the main pattern.

The actual process of colony division begins at mid-morning with the exodus of the queens. The functional queen and her entourage of workers leave the bivouac on a foraging trail adjacent to her pole of the bivouac. Her progress is slowed by the bidirectional flow of traffic in the column. After some hours, the queen and a now stable cluster of workers are probably no more than 30 meters from the bivouac. The virgin queens move out on one of the other principal trails; the first to do so is usually the first eclosed and the leading contender to be a functional queen. Secondary virgin queens on the trail are eventually sealed off by workers and then abandoned. By late afternoon or early evening, the two functional queens—that is, the old queen and the victorious new queen—and their respective clusters of workers are well separated from the bivouac on divergent trails.

The process of fission climaxes with two diverging emigrations of the bivouac's remaining workers. Although the bivouac was polarized before the division, the newly eclosed males and the worker larvae of a new unisexual brood are divided evenly between the daughter colonies. This halving of males and larvae appears to be random, with the males moving under their own power and the larvae being transported by workers to their respective daughter colonies. After the dual emigrations are completed during the night, a thin, bidirectional column of workers passes through the deserted statary bivouac site connecting the daughter colonies (Schneirla 1956). This tenuous column, a lingering umbilical cord of consternated workers, normally disappears late on the following day. One colony has now become two.

In *Neivamyrmex nigrescens,* as in *Eciton,* sexual brood production and colony fission are correlated with, if not stimulated by, the dry season (Schneirla 1961). A sexual brood in this species consists of about a thousand males and a few queens—that is, a brood population that is less than 3 percent of an all-worker brood. Also as in *Eciton,* the size of the sexual brood is reduced by workers feeding on the eggs (Schneirla 1961).

The larvae of *N. nigrescens* develop during a shortened 10-day nomadic phase, and latent fission in the succeeding statary phase is evidenced in an increasingly polarized colony—the queen and her affiliated workers in one half of the colony, the sexual brood and their associated workers in the other half. Overt fission does not begin until the eclosion of the sexual brood stimulates divergent emigrations. Queens emerge before males. The old queen and the victorious young queen depart, in that order, with their respective worker nuclei, as in *Eciton,* but they leave on successive nights and as particpants in full-scale emigrations. Although the young all-worker brood of the next cycle is divided equally between the two daughter colonies, all of the males accompany the young queen. In this respect, *N. nigrescens* is conspicuously different from *Eciton.*

Colony fission has not been observed in other species of New World army ants, although winged males have been found in *Labidus praedator* nests (Borgmeier 1955) and in a colony of *N. opacithorax* (Schneirla 1961).

Although similarities in the patterns of colony fission are evident in epigaeic species of Asian *Aenictus* and surface-active species of *Eciton* and *Neivamyrmex,* Schneirla (1971) considered the *Aenictus* pattern to be the most generalized. The most striking departure from the ecitonine pattern is the time at which *Aenictus* males take flight from the dividing colony. This aerial exodus occurs over several days and is completed before colony division rather than afterward. Compared with *Eciton,* the process of fission in *Aenictus* is compressed, with the exodus of queens and the divergent emigrations of the subdividing workers overlapping in time (Schneirla 1971).

Among the driver ants studied thus far—*Dorylus (Anomma) nigricans* and *D. (A.) wilverthi*—colony division is preceded by an early bipolar organization within the colony not unlike that discovered in *Eciton* (Raignier 1959, Leroux 1982). Division begins with the exodus of the mother queen and up to half of the workers and worker brood. The new colony and the entire sexual brood, including the yet-to-be-selected queen, remain in the old nest. The sexual brood may be in various stages of developement—the males may be larvae, pupae, or adults, and the queens may be callow or fully pigmented (Raignier 1959)—but once the functional queen exits from the nest with her half of the worker force, the daughter colony workers kill all but one of the virgin queens. Raignier (1959) found as many as fifty-six virgin queens in a single nest and discovered that the first queen to eclose is not necessarily the one spared in the elimination process. After the males eclose, they fly from the nest. In both *nigricans* and *wilverthi,* males are also produced in the absence of queen brood and do not necessarily presage colony division (Raignier 1959, Leroux 1982).

The mother colony in *D. (A.) nigricans* is characterized after the fission by its frequent, successive, and longer than usual emigrations, half of which result in nest site occupations of only one to four days (Leroux 1982). Leroux (1982) observed that colonies of this species increase in size by about 53 percent per year and undergo fission an average of every 22.6 months. Males, on the other hand, are generally produced every 12 months. In contrast with the correlation between colony fission and the advent of the dry season that Schneirla (1971) determined for *Eciton,* colony fission and the exclusive production of males in *D. (A.) nigricans* occur after the beginning of the rainy season (Leroux 1982). Leroux also found that males enclose two or three weeks after the queens, a much longer lag time than in *Eciton.*

Like much of what we know about army ants, our understanding of colony fission is limited to the conspicuous epigaeic species. Hypogaeic species do whatever they do in the impenetrable lividity of their private labyrinths.

Mating

The periodic but infrequent exodus of males from the army ant colony signals the beginning of a reproductive odyssey that ends in success for perhaps only a few individuals. To be reproductively successful, males must find alien conspecific colonies, and they must do so in the face of legions of hungry predators. Most often, it seems, they discover the foraging or emigrating columns of workers of such colonies and attempt to join the workers on their march to the nest (although they may also find old pheromone trails and follow them to the nest). But to join the column they must be accepted by the alien workers; that is, they must "break the worker barrier" in order to reach the alien queen (Franks and Hölldobler 1987, p. 238). The workers, in admitting or denying admission of a male to their nest, are in reality choosing the father of their future nestmates, just as they choose the mother of their nestmates at the time of colony division.

Franks and Hólldobler (1987) reasoned that the principles of sexual selection and female choice should apply to workers, which *should* choose mates for their queen because they, the workers, will invest their lives in the survivorship of any stepsisters produced. Franks and Hölldobler further noted that if workers are to maximize their own inclusive fitness, they should choose males that offer the greatest fertility.

Sexual selection theory suggests that males may indicate the quality of their genes and gene combinations through a variety of physical traits—and specifically by being rather robust—and Franks and Hölldobler (1987) hypothesized that army ant males superficially resemble the queen in size and shape because workers use similar criteria to select both queens and their mates. In other words, worker participation in sexual selection has favored males that are most like the queen in form.

In addition to this convergence in morphology, the glandular equipment of the male abdomen—and presumably its exocrine secretions (noted in Chapter 3)—is also similar to that of the queen. In the competition for acceptance by the alien workers and consequent access to their queen, only the most "attractive" males succeed; that is, only the males with the most queenlike physique and parfumerie can expect a nuptial reward. Franks and Hölldobler (1987, p. 241) concluded that

> sexual selection acting on both the competitive ability of males and queens and the discriminatory ability of workers had led to a convergence in the way in which both sexes demonstrate their potential fertility. We suggest that this is the reason why males have similar size, shape and morphology to conspecific queens and furthermore appear to release

attractive pheromones from the same sites on their bodies. Workers in army ants are intimately involved in all the selection procedures in reproductive colonies. Army ant colonies are therefore perhaps the most "democratic" of all insect societies.

As alluring as this hypothesis of male-queen convergence may be, I am not convinced that the male's robustness is convergent, or that the male's battery of queenlike exocrine glands is more than the result of homology. Perhaps the rigorous agenda the male army ant must fulfill to reach a mate precludes the existence of the petite male habitus common to other ants.

The time at which males fly from their colonies on their reproductive quest is not the same for all species. In the first nightly emigrations of newly formed *Eciton* colonies, the alate males stay in the line of workers, walking under their own power but subservient to the movement of the workers (see Figure 3.22). After the third and fourth nights, however, the males are less inclined to follow the line of march, and many depart from the emigration trail just outside the bivouac (Schneirla 1971). With each subsequent nightly migration, more males break away from the column and take flight. Most males probably leave emigrations in the middle of the nomadic phase.

Neivamyrmex nigrescens males also take flight following colony division. The newly eclosed males of the phasic species of *Aenictus*, on the other hand, fly from the colony before it divides (Schneirla 1971). In *Dorylus*, at least among the driver ants, the appearance of males flying from the nest does not necessarily mean that colony division is imminent. Sexual broods may be exclusively male (Raignier 1959, 1972; Leroux 1982). Raignier (1959) did observe colony fission when only male brood was present, but in these extraordinary cases the division was asymmetrical. While most of the colony emigrated with the functional queen, a few workers were left behind with the pupating males. Once these emerged and took flight, the abandoned, queenless workers died. The flight of males of *D. (A.) nigricans* in Ivory Coast occurs over a three- to five-day period, primarily at dusk and dawn (Leroux 1982). Initially, the restless young males of this species are restrained from flying by the workers.

Once a young male finds an alien conspecific column and is permitted by the workers to join them, it either sheds its wings or has them torn from its thorax by the workers. Although "the loss of wings further increases the superficial resemblance of males to queens, this trait," Franks and Hölldobler (1987, p. 239) speculated, "is probably also functionally related to the process of wing muscle histolysis which will provide a male with both energy and resources probably for spermato-genesis, the production of large quantities of attractive pheromones and sustenance for the relatively long period it must live with the new colony before it gains access to the queen." Although some of Franks and Höll-dobler's speculations may be correct, the authors are wrong about sper-matogenesis, which occurs during pupation (Gotwald and Burdette 1981).

Dealate males have been found running in the columns of *Eciton* (Schneirla 1971), *Labidus praedator* (Rettenmeyer 1963b), and *Dorylus* (Savage 1849, Leroux 1982, G. R. Cunningham–van Sommeren, pers. comm.). The first recorded observation of this phenomenon appears to be that of the medical missionary Dr. Thomas S. Savage (1849) for the genus *Dorylus*—and this before the workers and males had been rec-ognized as belonging to the same species.

In his field studies of *Dorylus nigricans,* Leroux interpreted the process of male acceptance by alien workers rather differently than Franks and Hölldobler (1987) saw it for the ecitonines. His observations suggested that the males are rather passive once they have come upon a column of conspecific workers, which seize the males, remove their wings, and take them back to the nest (Leroux 1982, p. 12). And Rettenmeyer (1963b, p. 414) observed a dealate male in a *Labidus praedator* emigration column "with so many workers clinging to it that it could scarcely walk."

Mating in army ants has seldom been observed. The mating pair of *Neivamyrmex carolinensis* discovered as a nest was being excavated is, perhaps, the only copulating pair ever seen in nature (M. R. Smith 1942). Schneirla (1949) twice observed mating in captive *E. hamatum,* and one *E. burchelli* mating was seen by Rettenmeyer (1963b).

The first *E. hamatum* mating that Schneirla observed took place be-tween a queen and one of two dealate males taken from a bivouac. On returning to the laboratory, Schneirla (1949, p. 45) found the two already in copula, the male mounted over the queen "with his mandibles hold-ing tightly to one of her large petiolar horns and the tip to his gaster so extensively inserted into her abdomen as to deform her gaster consid-erably." The male appeared lethargic throughout the two hours the pair remained coupled, but the more restive queen carried the male with her as she ran about the container (Schneirla 1949). The second mating also involved a queen and dealate male, these removed, with a large cluster of workers, from a bivouac. Again, they were already in copula by the time Schneirla reached the laboratory. The coupling lasted 10 hours. When dissected, the queen was found to have several large, discrete, ball-like masses of spermatozoa in her spermatheca. Mating in *Eciton,* Schneirla (1971, p. 260) concluded, "is a protracted event in which the male inserts maximally into the female, seeming literally to explode his sperms into her gaster, after which he dies." Bulmer (1983) suggested that this form of mating, in which the male mates once and dies, is a consequence of the excess of male over female reproductives produced by social insects that swarm. He postulated that, as mating opportunities are so rare, a male should risk everything when he has the opportunity to mate, in order to maximize his return. It should be noted, however, that the death of males following the single copulatory event of their lives is more the norm in insects than not, and army ant males are certainly not unique in this respect. A colorful description of male ants and their function was penned by Booker Prize–winning author A. S. Byatt (1992, p. 119): "They are flying amorous projectiles, truly no more than the burning arrows of the winged and blindfold god of love. And after their day of glory, they are unnecessary and unwanted."

In an act of myrmecological matchmaking, Rettenmeyer (1963b) placed together in a petri dish a queen and dealate male of *E. burchelli* taken from the same emigration column. Copulation started within one minute and persisted for an hour (Figure 4.16): "For the first 15 to 25 minutes the male constantly moved his antennae and front legs in large arcs. Only occasionally did it look as though he was stroking the queen with his antennae. The male held the queen by his middle and hind legs, never with his front legs. He also grasped her petiole behind the horns with his mandibles. At no time did he hold onto the horns themselves. The queen walked around a little during the copulation but was never very active" (Rettenmeyer 1963b, p. 373).

Although mating in *Aenictus* and *Dorylus* has not been observed, Raignier and van Boven (1955) conjectured that driver ant queens are fertilized many times during their lives. They also speculated that the driver ant queen may even be inseminated by males of different subgenera. On one occasion, for instance, a wingless male of the subgenus *Typhlopone* was seen among the workers of a colony of the driver ant *Dorylus wilverthi* (van Boven, personal communication). Such an interspecific association, of course, does not imply subsequent insemination.

Interspecific Interactions

Chance encounters between army ant colonies would seem inevitable, with foraging and emigrating workers crisscrossing the forest floor like so many automobiles speeding about on modern, interconnecting highways. A colony of *Dorylus* (*Anomma*) *molestus*, for example, was involved in five encounters with two other *molestus* colonies within a 432-day

Figure 4.16. A copulating male and queen of *Eciton burchelli*. These individuals were taken from the same emigration column and placed together in a laboratory petri dish. Copulation commenced within one minute and persisted for an hour. (Photograph by Carl W. Rettenmeyer, Connecticut Museum of Natural History. Reprinted, by permission of Carl W. Rettenmeyer and the University of Kansas, from Rettenmeyer 1963b.)

period (Gotwald and Cunningham–van Someren 1990), although Franks and Fletcher (1983) calculated that encounters between *Eciton burchelli* colonies on Barro Colorado Island occur once per colony every 250 days. As noted previously, these latter investigators found that *E. burchelli* colonies move in directions determined by the most immediate previous raid or emigration. Conspecific colonies on Barro Colorado Island, Franks and Fletcher (1983) concluded, exhibit mutual avoidance. In Africa, the driver ants, the ecological analogues of *Eciton,* also appear to avoid collisions, and agonistic interactions are infrequent. Leroux (1982, p. 142) submitted that aggression between *D. (A.) nigricans* colonies was rare. In only two instances did he observe *Anomma* colony collisions that resulted in battle. Asian *Aenictus* do not appear to attack congenerics, nor do they seem to attack other ants with army ant lifeways (e.g., *Leptogenys*) (Rosciszewski and Maschwitz 1994).

Encounters between colonies of different species, on the other hand, are not infrequent. Uneventful meetings between colonies of *E. burchelli* and *E. hamatum* are common on Barro Colorado Island (Franks and Fletcher 1983), and Schneirla (1949) reported that a raiding column of *E. hamatum* returning to its statary bivouac intruded rather significantly on the emigration of an *E. burchelli* colony when the columns of the two colonies crossed in two places. He recorded little or no aggression between the workers at the collision points. Schneirla (1949) also described an encounter between *E. hamatum* and a raiding column of *E. mexicanum*, in which the latter species bisected the path of the former species. Rettenmeyer (1963b) noted that while all the army ants he studied captured more ants than any other kind of prey, they never took other species of ecitonines.

Not all interspecific encounters are peaceful. Borgmeier (1955) saw *Nomamyrmex esenbecki* carry off workers of *Eciton dulcius*. On Barro Colorado, *N. esenbecki* was observed in small skirmishes with *Labidus coecus* and *L. praedator*, but in no instance did a worker of one species carry off a worker of the other (Rettenmeyer 1963b). Rettenmeyer wrote about one rather ritualized encounter: "When *esenbecki* had a raid column touching that of *coecus*, about ten workers of each species traded bites and vibrated antennae at each other for more than four hours. The same individual ants did not persist in this battle for the entire period, but the number of workers stayed relatively constant. Neither species attacked the other army ant as decisively as it attacked other insects near by, and no workers of either army ant appeared to have been injured" (1963b, p. 426).

The most dramatic interspecific collisions between army ant colonies involve *Dorylus* species of the subgenus *Typhlopone*. In 1979, in Ghana, my student Robert Schaefer and I found what could only be described as a battlefield—the graphic evidence of a violent encounter between *D. (T.) obscurior* and the driver ant *D. (A.) nigricans*. My field notes read: "Road littered with corpses of *Typhlopone* and *Anomma* workers. Many died, impaled, in the grasp of each other's mandibles; fragmented and dismembered bodies are conspicuous among the corpses that cover a triangular area of exposed, dirt road measuring 8 × 5 × 5 meters."

This observation is not unique. Leroux (1982) reported that *Typhlopone* attacks not only foraging columns of *Anomma* but its nests, as well. Of the nine attacks on *Anomma* nests that Leroux witnessed, six resulted in the destruction of the queen and her colony. Following such attacks, the demise of the colony is further ensured by the invasion of other predators, such as the ponerine ant *Paltothyreus tarsatus*. The coup de grace for the *Anomma* colony is often delivered in subsequent attacks launched by *Typhlopone*. Leroux (1982, p. 144) regarded *Typhlopone* workers as the most formidable and efficacious predators of *Anomma* colonies and believed that *Typhlopone* certainly plays a role in the population biology of *Anomma*.

5

Guests and Predators

A veritable menagerie of arthropods and vertebrates maintain associations of one sort or another with army ants. Although such associations are found elsewhere among the ants, the level of biotic pluralism achieved within and about army ant colonies is truly astounding. All these guest and associate species, numbering in the hundreds, perhaps thousands, exploit either the ants themselves or the nest environments they create. Add to this roster of associates the arthropods and vertebrates that include army ants on their menus as serendipitous fare, and the magnitude of the role of army ants, including their influential occupation as predators, in tropical and subtropical ecosystems becomes abundantly clear.

The role of army ant predators is narrowly defined and obvious; that of guests and associates is not. Indeed, the roles that guests and associates play are richly varied and fulfill myriad possibilities. For some species, their roles remain elusive and enigmatic.

Certainly most associated species are symbionts, organisms that maintain a coincident and dependent relationship with the army ants. Many symbionts are benign in their exploitation of the army ants. These commensals benefit from the relationship but neither harm nor help their army ant hosts. Some mites, for example, are simply phoretic, mere riders on the bodies of the ants, minute commuters traveling about the forest and savanna floor farther and more efficiently as hitchhikers than they could under their own power. Still other commensals feed on the refuse found in the middens, the garbage dumps of army ant nests (see Chapter 3).

Many species feed either on the prey of army ants or, as predators or parasites, on the army ants themselves. Obviously, then, some symbionts possess the key that unlocks the fortress door. Using covert means, they have gained entrance to a notoriously well-defended colony. Once through the portal, still others, consummate thespians, play an adaptive charade, pretending to be what they are not: members of the colony.

Symbionts and Guests

The symbionts, loosely defined here to include even species that are strictly phoretic on army ants, are profligate in the variety of their form and behavior. So diverse are they that they have rather successfully defied attempts of myrmecologists to categorize them in ways that reveal their biological modus operandi and their ecological place in the army ant scheme of things.

The most enduring attempt at categorization is the system devised by Erich Wasmann (1894), a Jesuit priest who pioneered the systematic study of ant guests. Wasmann created six categories of army ant symbionts: (1) synechtrans, guests that are perceived as aliens and are actively and aggressively pursued by the ants but that often elude their unwilling hosts by speed and agility or protect themselves beneath carapace-like bodies or with defensive chemicals; (2) synoeketes, arthropods that are simply tolerated (Wasmann 1912), ignored by their hosts because of their neutral odor, slow movements, or peculiar shape; (3) symphiles, or "true guests," which are accepted by the army ants and integrated into the social life of the colony; and (4) ectoparasites, arthropods that live on the body surfaces of the ants, ingesting secretions or exudates from the exoskeleton or hemolymph after piercing the integument. The other two categories, endoparasites and trophobionts, are not considered here, the former because nothing is known about the endoparasites of army ants and the latter because army ants are not known to regularly secure the honeydew excretions of homopterous insects. Among army ant guests and symbionts are some whose relationship with the ants is obligatory, at least for some part of their life cycle. These myrmecophiles are specifically adapted to coexistence with the ants and depend on them for survival. In spite of the fact that some myrmecophiles prey on the army ant brood, they are tolerated by their hosts—indeed, are commonly integrated into the social fabric of the colony (Kistner 1979, 1982).

Renaud Paulian (1948), who studied beetle guests of the West African species *Dorylus* (*Anomma*) *nigricans*, proposed a scheme that categorizes these beetles as clients, followers, or associates. Clients, he explained, feed on the detritus and cadavers in the middens or prey on other arthropods attracted to these refuse piles. The behavior of clients is not integrated in any way with the behavior of their hosts, with whom they have a detached relationship. Followers, on the other hand, are behaviorly synchronized with the activities of their hosts and are commonly found in their hosts' foraging and emigration columns. Then there are the associates, forms that have direct, active contact with the ants (Paulian 1948). Specifically, Paulian was referring to staphylinid beetles of the genus *Doryloxenus,* which ride on the backs of the workers, and consequently associates are phoretic. Although this system is useful in examining "dorylophiles," it lacks sufficient generality to be applied to the guests of the Ecitoninae (Kistner 1979).

Still another scheme, advanced by Cl. Delamare Deboutteville (1948), awkwardly rearranges some of Wasmann's categories into accidental commensals, incidentally present in the nest; preferential commensals, often present in the nest but not dependent on the ants; and obligatory commensals, which rely on their hosts for survival.

David H. Kistner (1979), a prolific investigator of myrmecophiles and termitophiles, produced the most recent scheme. A modification of criteria suggested by Akre and Rettenmeyer (1966), it emphasizes the degree of specialization in the symbionts. Akre and Rettenmeyer (1966) dealt specifically with Staphylinidae associated with New World army ants and devised a list of behavioral attributes that distinguish generalized species from specialized species. Generalized forms, they contended, are often found in foraging columns and live at the periphery of the nest or bivouac or in the middens, whereas specialized forms are uncommon in foraging columns and live within the nest or bivouac.

Kistner's (1979, pp. 343–344) version rates symbionts as integrated species, which are incorporated into the social organization of their host, or nonintegrated species, which are adapted to the ecological niche the nest provides but are not integrated into the colony's social milieu. Kistner (1966) also developed an integration index for calculating a symbiont's social status. The index is designed to indicate the extent of integration and is based primarily on the myrmecophile's location within an army ant column. If a species predictably travels in the central parts of columns, where traffic is congested, it is considered integrated; if, on the other hand, it travels at the ends of columns, it is deemed nonintegrated. The degree of integration is computed as a percentage of all individuals of a particular species found moving in the central portion of a column as opposed to other parts of the same column.

No system of symbiont functional classification yet devised is completely satisfactory. Wasmann's system essentially ignores certain symbionts—antbirds of the family Formicariidae, for instance—that may be obligatory followers of army ant foraging swarms. And the degree of specialization and integration of symbionts, vis-à-vis the approaches of Akre and Rettenmeyer and Kistner, are difficult to measure. I propose here a different approach that considers both the habitual or primary location of the symbiont or guest and the symbiont's major food source (Figure 5.1). I do not disregard the importance of evaluating and understanding the social integration of symbionts, but it is also important to be able to organize symbionts and guests into categories whose criteria are easily assessed empirically. Indeed, any generally useful system must also consider the degree of the symbiont's dependence on its host, an unalterably important ingredient in the examination of myrmecophilous relationships. Initially, it is logical to sort species into intranidal symbionts and guests (those found within the nest) and extranidal symbionts and associates (those living outside the nest) (Gotwald 1994).

Intranidal symbionts and guests are of four types: (1) facultative predators, which prey on army ant brood but do not depend on the army ant colony for their continued livelihood; (2) facultative and obligatory commensals, including phoretic species, which are preadapted to the nest environment and do not attack the ants; (3) obligatory symbionts, both predators and prey feeders, which depend on the ant colony for survival and are integrated, to some extent, into the social life of the colony; and (4) ectoparasites, most notably mites, which live on the exudates and hemolymph of the ants.

The extranidal symbionts and associates are, with the exception of those that live in the middens, camp followers that make their living, at least in part, by locating and following army ant columns and swarms and attacking arthropods flushed by the ants or the ants themselves. These symbionts and associates include facultative detritivores and scavengers that live in the middens; facultative predators of two types: those that prey on host ants and other arthropods that frequent the middens and those that locate army ant columns and attack the ants; and swarm followers, which exhibit varying degrees of dependence on the ants, including species that attack potential army ant prey scattered by the advancing tide of foraging ants. Among the swarm followers are the cleptobionts, a small group of species that steal prey directly from the mandibles of the worker ants, and the scatophagous butterflies, which pursue army ant swarms and thereby locate and feed on the excreta of swarm-following birds.

Abundance and Diversity of Guests and Associates

Although only a few quantitative studies of army ant guests and associates have been published, these make it eminently clear that army

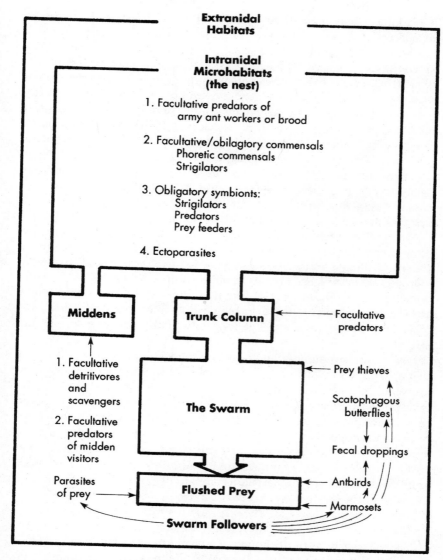

Figure 5.1. A functional classification of army ant symbionts, intranidal and extranidal, based on the habitual or primary location of the symbiont or guest and on the symbiont's major food source.

ant colonies are home to thousands of individual guests and a potpourri of species and genera, even orders and phyla. Noninsect arthropod guests and associates include a single species from the rare tropical arachnid order Ricinulei (Fage 1938), six or more species of spiders (Bruch 1923, Fage 1938), mites belonging to at least 12 families (Rettenmeyer 1960, 1962a), and three genera of millipedes (Rettenmeyer 1962b).

Insects abound among army ant guests and followers. The seven orders represented include four species of bristletails (thysanurans) (Rettenmeyer 1963a, Wygodzinsky 1982), springtails (collembolans) (Delamare-Deboutteville 1948), beetles belonging to five families, minute parasitic insects called strepsipterans (Kistner 1982), butterflies belonging to three families (Drummond 1976, Ray and Andrews 1980, Lamas 1983), flies from seven families (Kistner 1982), and parasitoid bethylid and diapriid wasps (Evans 1964, Masner 1977).

Among the vertebrate guests and associates are subterranean insectivorous blind snakes that follow army ant pheromone trails (Gotwald 1982) and birds, specifically those regarded as "professional" army ant followers, from at least 4 avian families (Willis and Oniki 1978). Additional species of birds representing more than 13 other families follow army ants occasionally to fairly regularly (Willis and Oniki 1978). No other ant colonies rival the heterogeneity of this array of guests and associates.

Most abundant among the guests are mites, a subclass of arthropods that approaches the insects in total number of species. In a survey of 150 army ant colonies, Rettenmeyer (1962a) collected from the nest or bivouac twice as many mites as myrmecophilous insects. In some colonies, the ratio of mites to insect guests was 100:1. With 126 described species of myrmecophilous mites in 33 genera, 21 of which are unique to army ants, these mites comprise a diverse fauna in their own right (Eickwort 1990). Three of the mite families are found only with New World army ants, constituting a singular fauna apparently unmatched in the Old World tropics (Kistner 1979, 1982). Myrmecophilous mites are either phoretic or ectoparasitic on the ants (Figures 5.2, 5.8, 5.13, 5.20).

Second to the mites and the most abundant of the insect myrmecophiles, at least with the Ecitoninae, are minute cyclorrhaphous flies of the family Phoridae (Rettenmeyer 1962a). Rettenmeyer and Akre (1968) estimated that two army ant colonies they observed, one of *Labidus praedator* and the other of *Eciton vagans,* each harbored in excess of 4000 phorids. More than 200 species of phorids have been collected with ants, and certainly the vast majority of the myrmecophilous forms in the New World are associated with army ants (see Figure 5.9). These species, most of which live with more than one host species, are scavengers in the

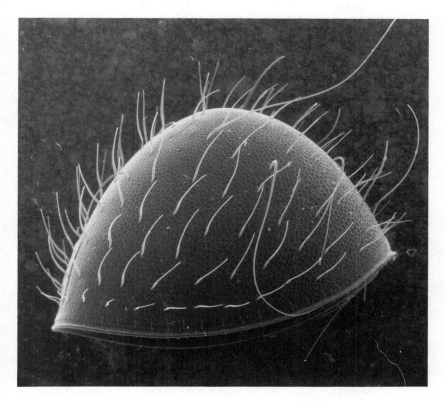

Figure 5.2. The phoretic mite *Circocylliba oligochaeta*, found on army ants of the genus *Eciton*, is approximately 1 mm long. (Photograph by Richard J. Elzinga. Reprinted, by permission of Richard J. Elzinga and the Muséum National d'Histoire Naturelle, from Elzinga and Rettenmeyer 1974.)

middens but may also feed on prey and army ant brood within the nest or bivouac (Rettenmeyer and Akre 1968).

Although myrmecophilous beetles of the family Staphylinidae may be represented in army ant colonies in smaller numbers than phorid flies—Akre and Rettenmeyer (1966) found fewer than one beetle per thousand ants in ecitonine colonies—the number of described species is greater than any other group of myrmecophiles. Aenictophilous, dorylophilous, and ecitophilous Staphylinidae of five subfamilies and more than 138 genera (Seevers 1965) have been classified. This level of diversity should not surprise us, for this family of largely predaceous beetles has more than 28,000 recorded species (Seevers 1965). The staphylinids occur rather ubiquitously throughout inhabitable terrestrial habitats, and most

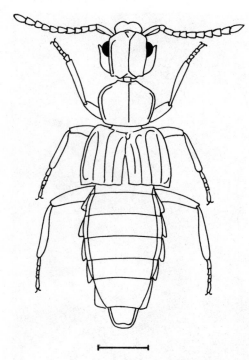

Figure 5.3. A generalized ecitophilous staphylinid beetle, *Ecitotropis carinata*. Scale equals 1 mm. (Redrawn from Borgmeier 1936.)

live in decaying organic debris where there is a ready source of food: larvae and soft-bodied arthropods.

Many staphylinids have evolved symbiotic relationships with termites and ants, relationships that often require some measure of social integration with their host species. Staphylinid beetles seem adaptively predisposed to myrmecophily, since they apparently comprise more than 20 independently evolved myrmecophilous groups, a polyphyletic assemblage, according to Charles H. Seevers (1965, p. 155), who produced an important monograph on the systematics, evolution, and zoogeography of these beetles. A majority of species, however, belong to a single subfamily, the Aleocharinae (Figure 5.3).

In Rettenmeyer's (1962a) survey of 150 New World army ant colonies, beetles of the family Limulodidae were the third most abundant guests. He collected 1100 limulodid specimens, impressively more than the 300 staphylinids collected. But this family of small, blind, wingless beetles is found only in the New World and Australia and contains but 30 species in eight genera (Dybas 1962), all found only with ants. Their streamlined, smooth, teardrop-shaped features probably protect the beetles

Figure 5.4. Beetles of the family Limulodidae, genus *Cephaloplectus*, with larvae of *Eciton rapax*. Their teardrop shape probably protects these beetles from attacks by their hosts. (Photograph by Carl W. Rettenmeyer, Connecticut Museum of Natural History.)

from attack by their host ants (Figures 5.4, 5.14). Limulodids feed on the cuticular exudates of the ant larvae, pupae, and adults (Dybas 1962).

The remaining taxa, ranked in order of abundance, were springtails, bristletails, millipedes, histerid beetles, and diapriid wasps (Rettenmeyer 1962a). A list of guests ranked by their total biomass might be even more helpful in understanding the impact of the colony's guests on its resources. Without a doubt, though, if we are to consider the worldwide abundance and diversity of army ant guests, three groups command our attention: mites, phorid flies, and staphylinid beetles. The majority of these and other arthropod guests are predators; detritivores and scavengers are next in abundance (Kistner 1982).

What accounts for the great diversity and abundance of army ant guests and associates? Field studies have revealed that the greatest diversity of myrmecophiles occurs with host species that form exceptionally large colonies. Army ants have huge colonies and numerous guests, as do non-army ants such as the meat ants of the genus *Iridomyrmex* (Wilson 1971a). "The insect colony and its immediate environment," E. O. Wilson (1971a, p. 391) wrote, "can be thought of as an island which symbiotic organisms are continuously attempting to colonize." Social insect colonies with the largest mature colony size, he noted, have the longest average span of mature colony life. That is, large colonies live longer. The issue is more complicated in army ants because new colonies are created by colony fission, which results in populations of related colonies akin to populations of cells—the parent cell of any one population is gone as an entity but continues on nevertheless as bits and pieces of future generations. So it is with army ant colonies, which never, perhaps, really die.

Long colony life, in turn, presupposes the high probability that symbiotic propagules will gain entrance to a given colony in due time. Furthermore, the equilibrium population size of symbionts will be proportionately large in large ant colonies, and their species extinction rate will consequently be low (Wilson 1971a). (Extinction in this case means the loss of symbiont populations from colonies [per colony per unit of time].) And finally, large colonies will have larger nests with a correspondingly larger number of microhabitats that can be colonized by a greater variety and number of guests. How many stowaways can a tugboat accommodate compared with an ocean liner? All the factors that result from large colony size—long colony life, low symbiont extinction rates, and high microhabitat heterogeneity—result in higher diversity of army ant guests and associates.

Integrating Mechanisms

The fact that so many arthropods have unlocked the fortress door, have penetrated the ant colony's elaborate outer barrier of chemical and mechanical recognition cues, implies that these arthropods have "attained the ability to speak the ants' language" (Hölldobler and Wilson 1990, p. 471). It is obvious that many speak the language well, since many are treated congenially by their hosts, some as if they were nestmates. This "language" and the other features that enable guests to insinuate themselves into the social organization of their hosts are called integrating mechanisms. They include, according to Kistner (1979), chemicals, external morphology (some ant guests look like their hosts, a phenomenon termed Wasmannian mimicry), and the signals by which the guests successfully solicit food from their hosts.

Chemical Integration

The use of chemicals by myrmecophiles to integrate themselves into the social organization of their hosts is no doubt common and of primary importance. These chemicals of appeasement and adoption, as Kistner (1979, p. 347) called them, sometimes originate from trichomes, tufts of golden hairs associated with glandular areas of the myrmecophile's integument. Secretions licked from these trichomes by the host ants have a calming effect, subverting the agitated aggressiveness that the ants normally direct toward interlopers. The tergal glands of certain myrmecophilous staphylinids are unquestionably another source of integrating chemicals (Kistner 1979).

But the use of integrating chemicals by aenictophilous, ecitophilous, and dorylophilous guests is not well documented; indeed, there is a disturbing paucity of information on the chemistry of appeasement and adoption among these myrmecophiles. Researchers have looked for such chemical integration, but their observations are inconclusive. Histerid beetles, for instance, that live with *Eciton burchelli* and *E. hamatum* possess what appear to be trichomes, but these clusters of hairs, located on the legs, thorax, and abdomen of the beetles, are not particularly attractive to the host ants (Akre 1968).

Other aspects of the exocrine chemistry of army ants and their guests are important to the symbionts' successful exploitation of army ants but do not qualify as integrating mechanisms (Kistner 1979). Critical to symbionts, for example, is their ability to detect and follow the chemical trails of army ants. Although ants may use subtle blends of exocrine chemicals to insulate their trails from ant species, they are not able to deter unrelated animals, including vertebrates, from using their trails (Blum 1974, p. 233).

Through trail following, myrmecophiles, especially flightless species, can find host colonies and also emigrate along with those colonies. In the latter case, the trail is a lifeline that enables guests to maintain a continuous relationship with a single host colony. Some myrmecophiles are so proficient at trail following that they can detect species-specific differences in army ant trails. They may be more sensitive to these differences than the ants themselves (Akre and Rettenmeyer 1968). Most myrmecophiles, however, seem to be limited to following freshly laid trails and trails whose persistence is not negatively affected by the type of substrate on which they are laid or by their exposure to rain (Torgerson and Akre 1970a). Some myrmecophiles enter new host colonies in a rather novel way, through an airborne phoretic transfer. Observing male driver ants attracted to a light one evening in Kenya, Kistner (1976) found that 1 in 20 males had a *Typhloponemys* beetle (Staphylinidae) walking about on its abdomen. It is tempting to imagine myrmecophiles moving from colony to colony on the bodies of army ant males like so many passengers on an airline shuttle service.

The acquisition by myrmecophiles of their host colony's odor may be important in their eventual integration into the host's society. In some ants, such as carpenter ants of the genus *Camponotus,* for instance, the queen produces odor labels, or discriminators, that are dispersed among, and learned by, the adult colony members. These acquired odor labels are membership cards for a rather exclusive club: the individual colony (Carlin and Hölldobler 1983). Obviously, no matter what the source of the discriminators, this identification system is not error free, and the club is therefore not completely exclusive. The importance of acquiring the colony odor in the process of being accepted by a host colony remains to be demonstrated for army ant symbionts. *Euxenister caroli,* an ecitophilous histerid beetle, possesses long hairs on the inner surfaces of its tibiae that may be involved in the transfer of colony odor from the host workers to itself. This beetle frequently rubs its tibial brushes over the body surface of a host ant and then over itself, possibly effecting the transfer (Akre 1968). Species-specific cuticular hydrocarbons have been identified in the ant genus *Solenopsis,* and the passive transfer of these hydrocarbons to a myrmecophilous beetle living with a colony of this genus has been demonstrated. The acquisition of these host hydrocarbons serves as an integrative mechanism for the beetle (Vander Meer and Wojcik 1982). The grooming of host ants by some staphylinids has also been interpreted as a way for ecitophilous guests to acquire colony odor (Akre and Rettenmeyer 1966, Akre and Torgerson 1968).

Wasmannian Mimicry

Although many ecitophilous and dorylophilous species differ little in morphology from their nonmyrmecophilous relatives, some species, especially staphylinid beetles, do resemble their host ants. Their myrmecoid body form is characterized by a narrowed, or petiolate, waist and an enlarged gasterlike abdomen (Figure 5.5). Additional myrmecoid features may include coloration, modifications to the thorax, integumental sculpturing, and antennal morphology (Seevers 1965, Kistner and Jacobson 1990) (Figure 5.6). The myrmecophilous Staphylinidae apparently experienced strong selective pressures to evolve the myrmecoid habitus, as evidenced by the number of groups that acquired this body form independently. Based on the structure of the petiolate waist in these beetles, Seevers (1965) estimated that there are at least 12 such groups.

The adaptive significance of this mimicry has been a matter for debate. Wasmann (1925) argued that the resemblance of ant guests to their hosts is in reality tactile mimicry because the host ants are often blind, or nearly so. Ants commonly touch or stroke nestmates with their antennae, perhaps for tactile recognition cues, and they do so with guests as well. Kistner (1979), who witnessed the interactions of staphylinids of the tribe Dorylomimini with their *Dorylus* hosts, noted that the ants palpated the

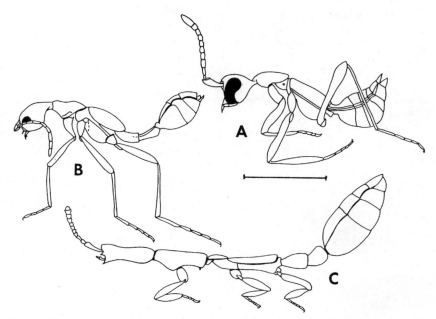

Figure 5.5. Three myrmecoid staphylinid beetles: *Parasahlbergius liberiae* (A; scale equals 1 mm), *Crematoxenus aenigma* (B), and *Mimanomma spectrum* (C). The myrmecoid body form is distinguished by a narrowed waist and an enlarged gasterlike abdomen. (Redrawn, by permission of the Field Museum of Natural History, from Seevers 1965.)

beetles with their antennae just as they would another ant. The beetles were never attacked or killed and always passed antennal scrutiny. This morphological masquerade, this "sheep-in-wolf's-clothing" adaptation, fools the host ant into accepting its inimical guest.

Because Erich Wasmann was the originator and leading proponent of the tactile mimicry concept, Rettenmeyer (1970) proposed that it be called Wasmannian mimicry. Although these myrmecoid mimics are often astonishingly like their hosts, others are less exacting impersonators. Duplicate copies of the ant hosts may not be required (Kistner 1979). Only the features that serve as social releasers need be morphologically counterfeited and available for tactile inspection. In mustering support for this view, Kistner noted that nonmimetic staphylinid guests of the genus *Typhloponemys* flex their abdomens over their backs when traveling with their Old World doryline hosts, thus creating the illusion of a gaster and petiole. The worker ants, he reported, palpate these flexed abdomens just as they do the gasters of their nestmates (Figure 5.7).

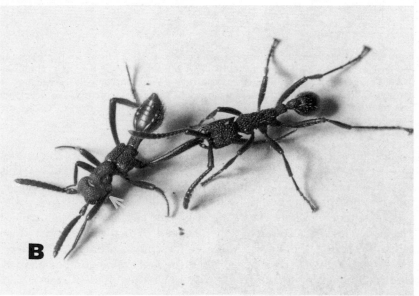

Figure 5.6. The myrmecoid staphylinid *Ecitosius robustus* (indicated by arrows) with its host, *Neivamyrmex sumichrasti*. Note how similar the beetle and its hosts are in body shape, integumental sculpturing, and antennal morphology. A. Beetle with host workers on a grasshopper femur. B. Staphylinid tugged at by an aggressive worker. (Photograph by Carl W. Rettenmeyer, Connecticut Museum of Natural History. Reprinted, by permission of Carl W. Rettenmeyer and the Kansas Entomological Society, from Akre and Rettenmeyer 1966.)

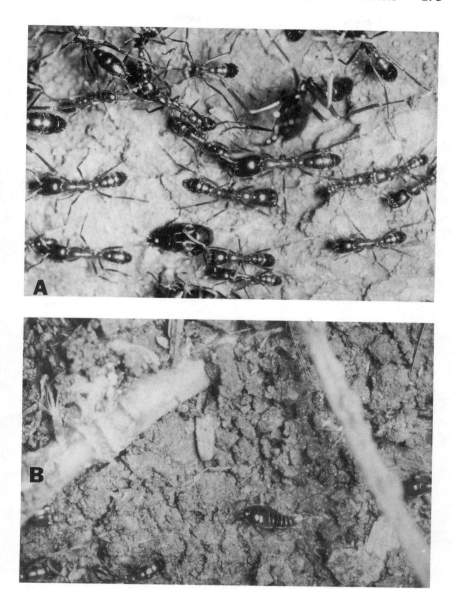

Figure 5.7. A beetle of the staphylinid genus *Typhloponemys* running in a column of its driver ant host, *Dorylus* (*Anomma*) sp. In the immediate presence of the ants, the beetle flexes its abdomen over its body, creating the illusion of a gaster and petiole (A), but in the absence of contact with the workers, it usually runs with its abdomen extended (B). (Photographs by David H. Kistner. Reprinted, by permission of David H. Kistner and Academic Press, from Kistner 1979.)

Some mimics even match the color of their hosts. The most dramatic example of this chromatic subterfuge is the staphylinid *Ecitomorpha nevermanni* (= *simulans*), which rather accurately matches the geographic color variations of its host, *Eciton burchelli*. In Panama and Costa Rica, the beetle approximates the reddish brown of its host; in Guatemala, it and its host are nearly black; and in Ecuador, both species are bicolored (Kistner 1979). (Kistner and Jacobson [1990] are now less enthusiastic about these color matches, since greater variation in color exists between beetles in the same host colony than was previously known.) Charles T. Brues (1902, 1904, p. 22) persuasively argued that the resemblance in color and form of myrmecophiles to their hosts when their hosts are blind or nearly so, as are the army ants, is "due to the influence of outside enemies." For instance, the color phenomenon may represent an adaptive response to vertebrate predators, namely, the ant-following birds that prey on the arthropods flushed by the foraging ants. The ants themselves are seldom taken by these birds, therefore looking like an army ant may keep the mimics from being devoured as well (Willis and Oniki 1978).

Because the majority of myrmecophiles that are integrated into the social organization of their hosts are not mimics, because mimics are most often found with surface-active species that are exposed to predators, and because social parasites rely on communication of one sort or another and chemical mimicry for acceptance by a host colony, Hölldobler and Wilson (1990) concluded that all mimicry that is visually apparent to taxonomists is probably directed at predators encountered outside the host colony. These authors, however, were unwilling to dismiss tactile mimicry entirely and conceded that some evidence suggests that tactile mimicry may also serve, in a supplemental way, the cause of social integration into the host colony.

Perhaps the most elegant example of Wasmannian mimicry is the phoretic mite *Planodiscus*. All species of this genus are exclusively phoretic on New World army ants and customarily attach themselves to the legs of their hosts, specifically to the undersides of the middle or hind tibiae (Elzinga and Rettenmeyer 1970, Elzinga 1991). Kistner (1979) used scanning electron microscopy to demonstrate that the sculpturing on the mite's body is essentially identical with that of the ant's leg. Furthermore, the arrangement and number of setae on the mite nearly duplicate the arrangement and number of setae on the ant's leg. Kistner (1979) concluded that the worker's tactile perception of the mite will be no different from its perception of its own leg, thereby ensuring that the mite will go undetected during grooming (Figure 5.8). In yet another example of Wasmannian mimicry, dermanyssoid mites of the newly de-

Figure 5.8. A mite of the genus *Planodiscus* on the leg of an *Eciton hamatum* worker. The surface sculpturing of the mite's body and the arrangement and number of setae on the mite are essentially identical with the ant's leg. (Photograph by David H. Kistner. Reprinted, by permission of David H. Kistner and Academic Press, from Kistner 1979.)

scribed family Larvamimidae (containing the single genus *Larvamima*) mimic the larvae of *Eciton*. The mites are likely to be carried during emigrations as if they were the larvae (Elzinga 1993). The selective forces responsible for the evolution of morphological mimicry in army ant symbionts may thus include the host ants themselves and their instinctive quest to positively identify nestmates, and extranidal predators that attack potential army ant prey but not the ants themselves.

Rogatory Behavior
The solicitation of food by guests from their ant hosts, functioning as an integrating mechanism, was dubbed rogatory behavior by Kistner (1979), who objected to the use of the word *begging* in descriptions of this phenomenon. He (p. 364) deemed *rogatory behavior* to be more neutral and to more accurately imply a ritualistic procedure. Furthermore, rogatory behavior can be seen as a form of Wasmannian mimicry in which social releasers are mimicked in order to exploit the behavior

of the ants. Kistner thus expanded the concept of Wasmannian mimicry to include symbionts' behavior as well as their morphology.

Rogatory behavior appears to be infrequently practiced by the guests of army ants, however, and understandably so because most myrmecophiles have remained predaceous throughout their evolution, and few have broken the rogatory code that would stimulate ants to feed them (Kistner 1982). Extensive laboratory observations of New World army ants and their staphylinid guests, for instance, have produced no conclusive evidence that ecitophilous staphylinids solicit alimentary fluids from their hosts in the mouth-to-mouth exchange of fluids commonly known as trophallaxis (Akre and Rettenmeyer 1966; Akre and Torgerson 1968, 1969). Only the histerid beetle *Euxenister caroli* has been observed to solicit and receive regurgitated fluid from its host, *Eciton burchelli,* and then only rarely (Akre 1968).

The Limuloid Body Form

The body forms of myrmecophilous arthropods, most notably beetles, provide convincing testimony that adaptive radiation has occurred within aenictophilous, ecitophilous, and dorylophilous guests and associates. Four morphological responses to the prevailing selective pressures for myrmecophily can be distinguished among symbionts, and each form includes species that are integrated to some extent into the social organization of their host colonies.

Most often, myrmecophiles closely resemble their nonmyrmecophilous relatives. Less commonly, myrmecophiles have acquired a myrmecoid habitus and resemble their hosts—with varying degrees of success as determined by the eyes of the biologist. Numerous other species, exemplified by the beetle family Limulodidae, possess a fusiform, or limuloid (cf. *Limulus,* the horseshoe crab) body form, characterized by a carapace-like covering and smooth teardrop shape. Lastly, there are specializations considered to be regressive, seen, for instance, in a number of phyletic lines of staphylinid beetles (Seevers 1965). These specializations include the loss of eyes or wings, reduction in the number of antennal and tarsal segments, and the fusion of various sclerites. The loss of eyes and wings is not correlated with body form and has occurred in generalized, myrmecoid, and limuloid myrmecophiles, but it is strongly correlated with the subterranean lifeways of the army ants (Seevers 1965).

The limuloid body form finds its prototype in the limulodid beetles (Figure 5.4). The entire dorsum of these beetles is smooth, streamlined, and basically seamless dorsally and anteriorly; the antennae are housed in

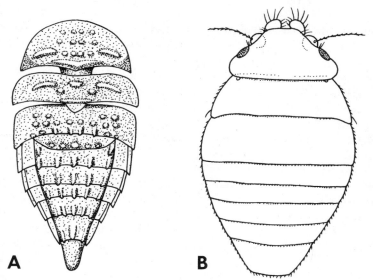

A   **B**

Figure 5.9. The limuloid body form in a staphylinid beetle, *Trilobitideus wasmanni* (A), and a phorid fly, *Aenigmatopoeus sequax* (B). In these dorsal views, the legs are not visible. The beetle is 2 mm. long, the fly 1 mm. (A redrawn, by permission of the Field Museum of Natural History, from Seevers 1965; B redrawn from Borgmeier 1963.)

deep furrows; and the heavily armed legs are short and retractable (Dybas 1962). This general body form is not uncommon among the Staphylinidae and can be found, for example, in the tribe Vatesini in the Neotropics and in some genera of the tribe Pygostenini and all the Trilobitideini in Africa (Seevers 1965) (Figures 5.9, 5.18a). The most extraordinary example of this body form is found in phorid flies that live with ants and termites, some of which look more like crustaceans than flies (Wilson 1971a). Less bizarre but still fusiform are members of the phorid genus *Aenigmatopoeus* that live in the middens of African driver ants. These flies also can be seen running in the raiding columns of the driver ants and at the ends of emigration columns (Kistner 1982) (Figure 5.9). But what is the adaptive significance of the limuloid body form? What selective pressures resulted in its independent emergence in such disparate groups?

Wasmann (1920) referred to this body adaptation as *trutztypus*, meaning "defense type," and proposed that the fusiform body protects its owners against attacks from host ants and termites, but *schutztypus*, "protection type," would be "more appropriate to its putative function," argued E. O. Wilson (1971a, p. 409). C. H. Seevers (1965) was of the opinion that Wasmann's explanation of this body form was oversimplified. Many limuloid species are too well integrated with their host colonies, Seevers reasoned, to depend solely on form for protection. Yet some observations support Wasmann's view. Fusiform staphylinids of the genus *Vatesus* in laboratory nests, for example, bury themselves in soil with their smooth backs exposed. From this position they resist with impunity attacks by ants (Akre and Torgerson 1969). In the field, Kistner (1976) observed individuals of the staphylinid genus *Anommatoxenus* slip from the clutches of driver ants. He (1976, p. 180) recorded an attempt by a medium-sized worker ant to grasp an *Anommatoxenus*. As the worker's "mandibles closed, the *Anommatoxenus*, with its limuloid shape and slippery surface, simply slid between them, head end out first." Whether defense is the dominant explanation or not, surely the convergent evolution of the limuloid body form in unrelated myrmecophilous taxa is proof positive of its adaptive importance within the army ant colony and nest.

Intranidal Symbionts and Guests

What can it be like within the army ant nest or bivouac? To be sure, the blind or nearly blind worker ants are rubbing shoulders with a multitude of myrmecophiles, many of which are nefarious gate-crashers involved in the intrigue of stealing prey or, worse yet, devouring their hosts. So talented are some of these guests at imitating their hosts that they are treated with sibling hospitality. Other guests are accommodating boarders that dine on leftovers or glean whatever secretions they can from the body surfaces of their hosts. And in this atmosphere of competing myrmecophilous agendas, the society's bona fide members go about the daily business of promoting colony growth and survival, no small task by any biological standard. Below are natural histories of selected myrmecophiles presented within the organizational framework that divides these symbionts, guests, and associates into intranidal and extranidal species. This is not an exhaustive survey of aenictophiles, ecitophiles, and dorylophiles—which are multifarious, to say the least— but I hope it reveals the complexity of the army ant's interspecific and interordinal relationships.

Facultative Predators

In the Usambara Mountains of Tanzania, the local people recount a tale of a snake they call *mkonko*, which is, they insist, the cow of the driver ants (Loveridge 1949). The snake they refer to is a fossorial blind snake of the genus *Typhlops,* a pantropical and subtropical genus of the family Typhlopidae. The widespread story describes an occasional *mkonko* slithering along in a column of driver ants, unmolested by the workers and protected by the soldiers. The ants keep their "cow" until they are short of food, at which time they kill and eat their domesticated guest. Given the fact that such snakes are known to feed on ants and termites, who eats whom after the snake enters the ant nest is hardly a matter for debate. Although I have never observed these snakes in *Anomma* columns, I unearthed one specimen when excavating a driver ant nest in Ghana (W. H. Gotwald, Jr., unpubl. data).

Similar subterranean blind snakes of the genus *Leptotyphlops,* family Leptotyphlopidae, are known to travel in New World army ant columns and follow army ant trails (Watkins et al. 1967, 1972; Gehlbach et al. 1971). *L. dulcis* was initially discovered in the nocturnal raiding columns of *Neivamyrmex nigrescens* and was not seen to eat the ants or the prey they carried back to the nest. Not only can this snake detect and follow the pheromone trail of the ant, it also seems to recognize a directional cue that guides it toward, rather than away from, the nest (Watkins et al. 1967). Watkins et al. (1967) hypothesized that these snakes follow foraging columns, trails, or both to the army ant nest, where they feed on the brood or prey. Laboratory studies demonstrated that *L. dulcis* can follow with equal skill the trails of termites and non-army ants as well as those of army ants. F. R. Gehlbach et al. (1971) assumed that this ability expands the spatial and feeding niches of blind snakes. It also strongly suggests that the snakes are facultative predators of army ants.

But how do these ophidian predators manage to crawl unmolested among the army ants? In the laboratory, and presumably also in the field, blind snakes *are* attacked by their "hosts." They respond by writhing and coiling, during which the tail moves rapidly between the coils and over the body surface. Gehlbach et al. (1968) observed that the writhing lasted about 0.33–2.50 minutes and was followed by a tighter, stationary coiled position that lasted 0.10–5.00 minutes (Figure 5.10), after which the blind snake returned to searching and feeding for 3 to 30 minutes without further ant attacks. During the writhing phase of its defensive behavior, the blind snake becomes covered with an odoriferous mixture of feces and a clear, viscous liquid that it discharges from its cloaca. Laboratory experiments established that this mixture repels the ants *Labidus coecus* and *Neivamyrmex nigrescens* (Gehlbach et al. 1968). It may also prevent the attacking workers from biting the snakes or grasping them in order to sting. Chemical analysis of the cloacal secretion of *L. dulcis* revealed that it is composed largely of a mucuslike glycoprotein containing a suspension of free fatty acids (Blum et al. 1971).

Figure 5.10. The blind snake *Leptotyphlops dulcis*, a predator of army ant brood, in the stationary coiled position that characteristically follows an attack by army ant workers. The snake is 183 mm long. (Photograph by F. R. Gehlbach. Reprinted, by permission of F. R. Gehlbach and the American Institute of Biological Sciences, from Gehlbach et al. 1968, *BioScience* 18:784–785, copyright 1968 by the American Institute of Biological Sciences.)

While they are writhing and coiling, the snakes also undergo a color change, becoming silvery. The change is probably produced by a lateral displacement of various scale rows and the tilting of individual scales at right angles to the body's longitudinal axis (Gehlbach et al. 1968).

Watkins et al. (1969) discovered that the cloacal secretion of *L. dulcis*, the only insectivorous snake found with *N. nigrescens,* repulses not only army ants but also other insectivorous snakes and snake-eating snakes. At the same time, this secretion acts as a pheromone to attract conspecific blind snakes, perhaps facilitating reproduction and the location of appropriate microhabitats. Watkins et al. (1969) concluded that repelling predators and competitors may be more important to the blind snakes than repelling ants.

Facultative and Obligatory Commensals
If any symbionts could be described as gentle as they go about their business, it would be the 10 or more New World species of myrme-

cophilous millipedes. Millipedes, of the class Diplopoda, are elongate arthropods with many pairs of legs that are primarily scavengers of decaying plant material. Found only with the army ant genera *Labidus* and *Nomamyrmex,* the myrmecophilous diplopods belong to three genera, the most important of which is *Calymmodesmus* (Rettenmeyer 1962b) (Figure 5.11). Because specimens survive in the laboratory, in the absence of army ants, for at least two months, longer than other isolated myrmecophiles, it is safe to assume that these animals are truly facultative visitors (Rettenmeyer 1962b). *Calymmodesmus* runs in the center of the emigration column (Figure 5.11), not at the edges like some symbionts, and is occasionally picked up and carried by the faster-moving worker ants. Rettenmeyer (1962b) reported from field observations that the millipedes are quite capable of following their hosts chemical trails, although laboratory experiments with *C. montanus* disclosed that they were not skilled trail followers and showed little or no preference for the trails of their hosts (Akre and Rettenmeyer 1968). Rettenmeyer also (1962b) noted that although *Calymmodesmus* species ate neither fecal material nor prey refuse of *Nomamyrmex,* they may help keep the ant nests clean. During the statary phase of about three weeks, perhaps much longer in the case of *Labidus* and *Nomamyrmex,* the two host genera, scavengers could help prevent the development of harmful mold or fungi by ingesting dirt around the bivouac.

At least five species of bristletails (order Thysanura, family Nicoletiidae) have been found with army ants, one species with the driver ant *Dorylus (Anomma) kohli* (Kistner 1982) and the others with ecitonines (Rettenmeyer 1963a, Wygodzinsky 1982). Thysanurans are primitive, wingless insects that characteristically possess three filamentous appendages protruding, tail-like, from the posterior end of the abdomen. Although it is obvious that one species, *Grassiella rettenmeyeri,* is truly facultative, for it can be found free living in forest leaf litter, the majority of species (belonging to the genus *Trichatelura*) may depend on their hosts for survival (Rettenmeyer 1963a).

Thysanuran myrmecophiles do not, however, appear to be host specific. *T. manni,* the most common species found with New World army ants, has been observed with at least six species of *Eciton* and one of *Labidus* (Torgerson and Akre 1969). This thysanuran lives within the bivouac and may prefer running among the central mass of larvae, a more tranquil part of the bivouac's internal structure. In the laboratory, *T. manni* fed on fluids oozing from pieces of prey. They were never seen to puncture or feed on the army ant larvae and pupae, but they did appear to feed on secretions and particles gleaned from the surfaces of the larval, pupal, and adult army ants and prey (Rettenmeyer 1963a).

Figure 5.11. Seven species of the millipede genus *Calymmodesmus* have been found with ecitonine army ants. Shown here is an individual running in an emigration column of *Labidus praedator* (A) and three individuals in a laboratory nest (B). (Photographs by Carl W. Rettenmeyer, Connecticut Museum of Natural History. B reprinted, by permission of Carl W. Rettenmeyer and the Kansas Entomological Society, from Rettenmeyer 1962b.)

They commonly climb over the workers, cleaning their bodies and legs, a behavior referred to as strigilation, a term originally coined by W. M. Wheeler (1910) (Figure 5.12). Rettenmeyer (1963a) speculated that bristletails may help keep the bodies and nests of the ants clean.

Ecitophilous thysanurans are most frequently seen in the center of the emigration columns of their hosts at times of maximum ant traffic when brood are being transported (Figure 5.12) (Rettenmeyer 1963a, Torgerson and Akre 1969). They do follow army ant trails but show no preference for the trails created by their hosts (Akre and Rettenmeyer 1968). In laboratory studies, Torgerson and Akre (1969) established that *T. manni* could sense ant trails through a number of receptors on the maxillary and labial palpi and antennae. They hypothesized that this would be quite adaptive to a vulnerable, soft-bodied guest that, in its jolting coexistence with the ants, had a propensity to lose some of its appendages. Akre and Rettenmeyer (1968) further concluded that because *T. manni* has a wide host range, individuals developing from eggs laid at times other than the statary phase may have to locate their hosts through fortuitous encounters with raiding or emigration columns.

Prominent among the intranidal commensals are phoretic mites belonging to various families, including the Cirocyllibanidae and Coxequexomidae, which are found exclusively with the New World army ants (Kistner 1982). Mites (class Arachnida, subclass Acari [Acarina]) have a tangled taxonomic history and are currently organized into a perplexing set of cohorts, orders, and suborders (Woolley 1988). The literature on aenictophilous, ecitophilous, and dorylophilous mites does little to clarify the phyletic relationships of these abundant myrmecophiles, and although numerous species have been described (see, e.g., Elzinga and Rettenmeyer 1966, 1970, 1974; Mahunka 1977a,b, 1978, 1979; Elzinga 1981, 1989), their biology eludes us beyond the fact that they securely attach themselves primarily to the workers of most, if not all, army ant species.

Phoretic mites found only with army ants have evolved a variety of holdfast mechanisms that permit them to cling effectively to their constantly moving hosts (Elzinga 1978). Species of the genus *Circocylliba*, for instance, use well-developed toothed claws to grasp the setae of their hosts. These species are also remarkably concave ventrally, which permits them to ride closely appressed to the broad curvatures of the host worker's head, thorax, or gaster (Figure 5.13) (Elzinga and Rettenmeyer 1974). Adult mites of the genus *Antennequesoma* attach to the host worker's antenna or leg by means of cuticular ridges in their uniquely formed lateral body flanges (Elzinga 1978). These species and other specialized phoretic mites are adapted to inhabit specific parts of the host, making

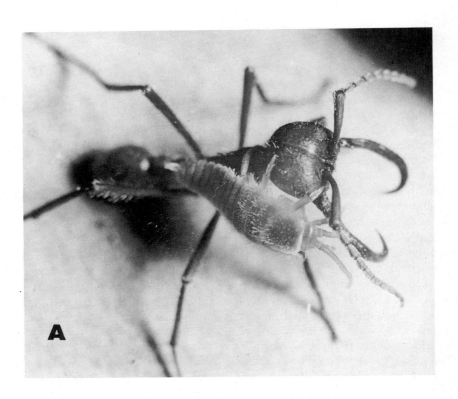

of each host an island composed of a variety of topographical micro-habitats.

Also reported from the nest of one army ant species, *Labidus coecus*, is the midge *Forcipomyia brumalis*, a nematocerous fly. Whether this apparent scavenger is a true myrmecophile remains open to question (Long 1902, Kistner 1982).

Obligatory Symbionts

Little is known of the biology of the myrmecophilous beetle family Limulodidae, discussed earlier in this chapter, except that their relationship with their hosts is definitely obligatory. Limulodids have been collected with *Eciton*, *Neivamyrmex*, and *Labidus* (Akre and Rettenmeyer 1968). *Paralimulodes wasmanni*, the sole species in its genus, is found with *N. carolinensis* and *N. nigrescens* and appears to be specific to *Neivamyrmex* (Wilson et al. 1954) (Figure 5.14). These beetles can be transferred in the laboratory from *N. carolinense* to *N. nigrescens*, but when placed in nests containing nonecitonine species they ignore the ants and wander about aimlessly (Wilson et al. 1954).

Limulodids only rarely run on the ground in the columns of their hosts, for they prefer to ride, usually on the heads and abdomens of workers, sometimes on queens (Figure 5.15). When two *Paralimulodes* beetles ride the same worker, they position themselves symmetrically on each side of the head or abdomen, similar to the way some mites ride their hosts (Wilson et al. 1954). In laboratory experiments, limulodids followed the chemical trails of species from the three host genera equally well (Akre and Rettenmeyer 1968). In crowded ant nests, *P. wasmanni* travel "over the bodies of their hosts in light, rapid, jerky movements, climbing appendages, scurrying around curved body surfaces, and skipping freely from one individual to another," apparently without disturbing their hosts (Wilson et al. 1954, p. 160).

Like other limulodids, *Paralimulodes* is a strigilator—it secures its food by scraping organic matter from the body surfaces of its hosts. Observations of *Limulodes paradoxus* in a laboratory nest of a non-army ant (genus *Aphaenogaster*) showed that these beetles prefer to strigilate the

---

Figure 5.12. Ecitophilous bristletails, in this case *Trichatelura manni*, commonly climb over their host workers, feeding on secretions and particles gleaned from their body surfaces (A). These thysanurans are most frequently seen running in the center of their hosts' (*Eciton rapax*) emigration columns (B). (Photographs by Carl W. Rettenmeyer, Connecticut Museum of Natural History. A reprinted, by permission of Carl W. Rettenmeyer and the Entomological Society of America, from Rettenmeyer 1963a.)

Figure 5.13. Phoretic mites, genus *Circocylliba*, attached to the mandibles of soldiers of *Eciton mexicanum* (A) and *E. dulcius* (B). *Circocylliba* mites are remarkably concave ventrally, which permits them to ride closely appressed to their host's body surface. (Photographs by Carl W. Rettenmeyer, Connecticut Museum of Natural History. B reprinted, by permission of Carl W. Rettenmeyer, from Rettenmeyer 1960.)

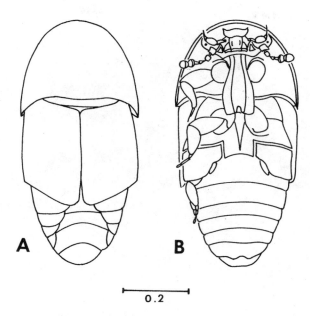

A   **B**

0.2

Figure 5.14. The myrmecophilous beetle *Paralimulodes wasmanni*, family Limulod-idae, shown dorsally (A) and ventrally (left legs omitted) (B), appears to be host specific for *Neivamyrmex*. Scale in millimeters. (Redrawn, by permission of E. O. Wilson, from Wilson et al. 1954.)

brood and clean host workers' integument only when brood are not available (Park 1933). Orlando Park (1933) noted that the mouthparts of *L. paradoxus* move back and forth in an arc as it feeds on the ant's body surface. Integumental exudates and foreign material are swept up by feeding brushes and conveyed into the mouth by toothed laciniae and inconspicuous, sickle-shaped mandibles. As meager as this biological information is, it seems enough to allow the postulation that the ecito-philous limulodids are obligatory commensals.

Beetles of the family Histeridae are commonly found with army ants; the genera *Euxenister* and *Pulvinister* with *Eciton*, for instance, and the genera *Coelocraera*, *Paratropus*, *Hypocacculus*, *Acritus*, and *Tribalus* with *Dorylus* (Akre 1968, Kistner 1982, Degallier 1983). Except for *Hypocac-culus*, these genera belong to the subfamily Hetaeriinae (Kistner 1982). Histerids are small, glossy black, usually oval beetles with truncated elytra (Figure 5.16). Their appendages can be tightly withdrawn into shallow grooves on the ventral surface of the body, giving them a smooth, inert appearance, a morphological attribute that preadapts them to myrmecophily. Found in dung, fungi, and carrion, histerid larvae and

Figure 5.15. Myrmecophilous beetles of the family Limulodidae on a *Nomamyrmex esenbecki* queen. (Photograph by Carl W. Rettenmeyer, Connecticut Museum of Natural History.)

adults prey on other insects that frequent these same microhabitats. New World species thus far studied are host specific and can follow the chemical trails of their hosts (Akre and Torgerson 1968). *Euxenister caroli* and *E. wheeleri* (Figure 5.16), however, when released on host trails that are 12 to 24 hours old, are unable to use them, suggesting that old chemical trails are not a host-finding mechanism (Torgerson and Akre 1970a). The same two species have functional wings and could locate their hosts by odor and fly to the bivouacs.

Although ecitophilous histerids can be seen running in the emigration columns of their hosts, most are phoretic and ride either the workers or the brood and prey carried by the workers. *Pulvinister nevermanni*, which lives only with *Eciton hamatum*, typically rides on the undersurfaces of the major workers' heads, often between their mandibles. *Euxenister caroli* and *E. wheeleri* in laboratory nests frequently climbed onto the workers (Figure 5.17), grooming them and at the same time rubbing their tibial brushes over the ants' bodies and then over themselves. The function of the grooming behavior, other than perhaps as a means of acquiring the host colony's odor, is not known; it may have a tranquillizing effect on the ants (Akre 1968).

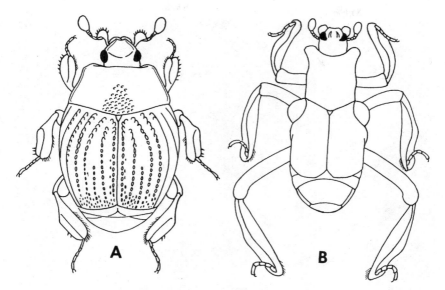

Figure 5.16. Ecitophilous beetles of the genera *Synodites* (A) and *Euxenister* (B). Histerid beetles are frequently found with army ants, but their biology is poorly known. (Redrawn from Bruch 1923, and Mann 1925.)

In laboratory nests, histerids frequently and efficiently cut open and ate the ant brood and prey (Akre 1968). The extent to which these species are integrated into the host colony or are simply tolerated guests has not been demonstrated, but their host specificity and the fact that they appear to be in reproductive synchrony with their hosts (Torgerson and Akre 1970a) suggests that their relationship is more obligatory than facultative.

The biology of dorylophilous histerids is poorly known. Kistner (1982) deduced that *Coelocraera* species are not host specific and not integrated into their host army ant societies. On the other hand, he has observed driver ant workers in the field carrying histerids, which suggests at least some level of communication or integration. I have also seen, on numerous occasions, histerids carried in the mandibles of West African driver ants in emigration columns (W. H. Gotwald, Jr., unpubl. data).

Many staphylinid beetles—generalized, limuloid, and myrmecoid—live with army ants and are integrated to some extent into the social lives of their hosts. These beetles feed either on their hosts or on host-captured prey. Species of the genus *Vatesus* live within the colonies of their *Eciton* hosts but are not tolerated. In laboratory nests, the adults of these limuloid staphylinids remained near the brood or prey, and al-

Figure 5.17. Histerid beetle, genus *Euxenister*, on the thorax of an *Eciton hamatum* soldier. (Photograph by Carl W. Rettenmeyer, Connecticut Museum of Natural History.)

though they fed on prey and dead brood, they were never observed killing live brood. *Vatesus* larvae run in emigration columns (as do adults) (Figure 5.18), and in laboratory nests they fed on live brood and the myrmecophilous thysanuran *Trichatelura manni* (Akre and Torgerson 1969).

Diploeciton is a myrmecoid staphylinid that runs in the emigration columns of its host, *Neivamyrmex pilosus*. In laboratory nests it spent all its time grooming worker ants of the same or smaller size. When grooming, the beetles assumed a rather unusual sidesaddle position on the host worker (Figure 5.19). Although they have been seen eating their host's brood, they may do so only when prey is not available (Akre and Torgerson 1968).

Ecitodonia setigera, a generalized staphylinid, can be seen in the emigration and raiding columns and in the middens of its host, *E. burchelli*. In emigration columns it moves at the same time the brood are carried

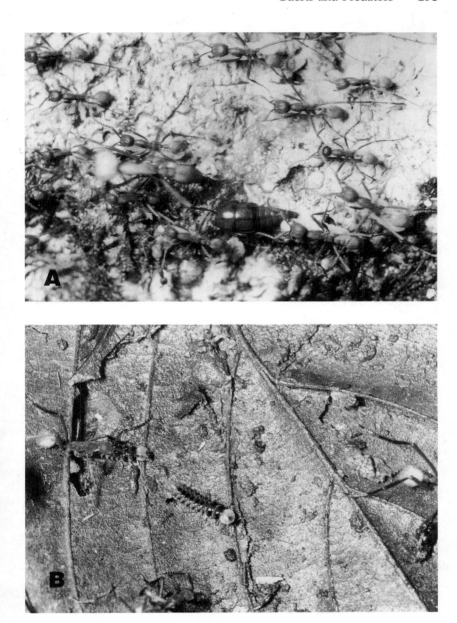

Figure 5.18. Adults (A) and larvae (B) of the limuloid staphylinid genus *Vatesus* run in emigration columns of their hosts—in this case, *Eciton burchelli* (A) and *E. rapax* (B). (Photographs by Carl W. Rettenmeyer, Connecticut Museum of Natural History.)

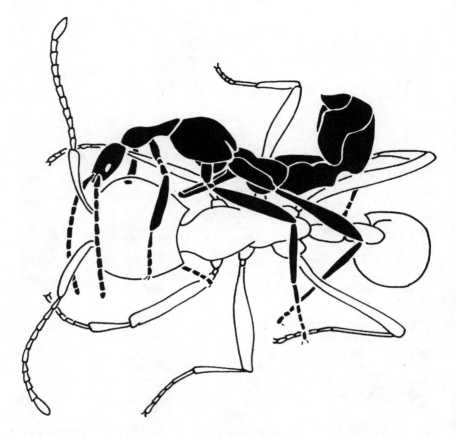

Figure 5.19. *Diploeciton nevermanni,* a myrmecoid staphylinid, grooming a worker ant. Note the beetle's sidesaddle position on its host. (Redrawn, by permission of R. D. Akre, from Akre and Torgerson 1968.)

and is ignored by the ants. These facts speak to its high level of integration. In laboratory nests, this species readily ate lycosid spiders and wasp larvae taken from foraging *E. burchelli* workers (Akre and Rettenmeyer 1966).

Our knowledge of the habits of Old World aenictophilous and dorylophilous staphylinids is limited. Laboratory observations are nonexistent, and most biological descriptions are restricted to the movement of these beetles in emigration and foraging columns. Little is known of their feeding habits. The taxonomy and phylogeny of these forms, on the other hand, have been explored in great detail by Kistner and his associates (Kistner 1966a,b; Jacobson and Kistner 1975a,b, 1981, 1983; Kistner and Jacobson 1975a,b, 1981, 1982, and many others).

Among the guests of *Aenictus* are the genera *Procantonnetia* and *Mimaenictus*; the former boasts a generalized habitus, the latter is myrmecoid. In emigration columns of their Asian host, *A. laeviceps*, these beetles are carried like larvae, slung beneath the bodies of the workers, which cradle the beetles by grasping the enlarged first segments of their antennae (Kistner and Jacobson 1975a).

Staphylinids of many types reside with African driver ants. Those with a generalized body form include the genus *Myrmechusa*, considered a nonintegrated, obligate predator on its hosts (Kistner and Jacobson 1982). Another genus of generalized form, *Dorylopora*, has been collected from driver ant middens, and its species are regarded as obligate scavengers that are integrated into their host colonies to varying degrees (Kistner 1966a).

Species of the myrmecoid tribe Dorylomimini, apparent Wasmannian mimics, are frequently antennated by their blind driver ant hosts when encountered in the ant columns. In this cautious greeting, the antennae of the inquisitor linger at the petiolate waist of the beetle; when the ant is satisfied that it has identified a sister worker, both move on (Kistner 1966b). Finally, dorylophilous species of the tribe Pygostenini are generalized or limuloid in form, are well integrated with their driver ant hosts, and appear to maintain an indiscriminate diet of prey and host adults, eggs, and brood (Kistner 1976).

New World wasps of the family Diapriidae, superfamily Proctotrupoidea, are not unusual in the columns of ecitonines, especially *Neivamyrmex* (Mann 1912, 1923; Borgmeier 1939; Huggert and Masner 1983). Proctotrupoids are parasitoids that attack the immature stages of other insects. Those found with army ants are myrmecoid and ostensibly mimic the ants both ecologically and morphologically (Masner 1976). Although winged in their dispersal stage, once these ecitophilous diapriids are associated with their host ants, their wings are spontaneously cast off or, perhaps, bitten off by the host ants. The wasps seem well integrated into the host colonies and are commonly carried by their hosts in raiding columns. Wasps of one genus, *Notoxopria*, conveniently possess a forward-projecting thoracic horn that functions as a handle which the ants clutch when carrying them (Figure 5.20). Kistner and Davis (1989) theorized that glandular exudates associated with this peculiar horn elicit the carrying behavior. In the absence of any data to the contrary, it is assumed that these wasps parasitize ecitonine larvae (Masner 1977).

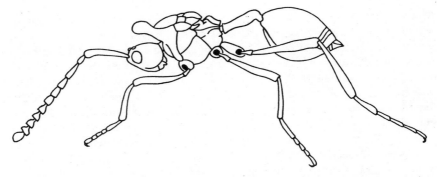

Figure 5.20. The ecitophilous wasp *Notoxopria cornuta*. Note the forward-projecting thoracic horn, which functions as a handle that the ants clutch when carrying the wasps. (Redrawn from Mann 1923.)

Ectoparasites

Although some species of the mite family Macrochelidae inhabit the middens of New World army ants and are not considered myrmecophiles, three species live as ectoparasites within colonies (Rettenmeyer 1962c). These host-specific species, uniquely adapted to their ectoparasitic roles, are exemplified by *Macrocheles rettenmeyeri,* which is found only with *Eciton dulcius.* Mites of this species insert their chelicerae into the membrane of the hind leg pulvilli (pads located at the bases of the tarsal claws) of the larger workers and presumably feed there (Figure 5.21). A bizarre and fascinating behavior is sometimes seen in bivouacs when the workers form nest clusters by hanging suspended from one another's tarsal claws. If a worker employs its mite-bearing leg in this process, the mite's curved hind legs substitute for the worker's claws. According to Rettenmeyer (1962c), the behavior of the ant appears unaffected when the hind legs of the mite serve as surrogate tarsal claws. Other mite ectoparasites live on army ants, but probably none can match the macrochelids for their adaptive ingenuity.

Orphaned in this brief inventory of animal ectoparasites is a single sporodochial fungus, *Termitariopsis cavernosa,* found thus far only on *Neivamyrmex* (Blackwell et al. 1980). This fungus, described as new from specimens of *N. opacithorax* and *N. pilosus,* is similar to fungi found on termites and some flies. The prevalence and importance of ectoparasitic fungi in army ants is unknown.

Figure 5.21. The ectoparasitic mite *Macrocheles rettenmeyeri* attached to the membrane of the hind leg pulvilli of an *Eciton dulcius* soldier. (Photograph by Carl W. Rettenmeyer, Connecticut Museum of Natural History. Reprinted, by permission of Carl W. Rettenmeyer, from Rettenmeyer 1960.)

Extranidal Symbionts and Associates

When army ants venture outside the nest, either on short refuse trips to the middens or on more protracted foraging and emigration expeditions, they are greeted by yet another throng of symbionts and associates. In general, these creatures are more often facultative than obligatory associates; certainly they are less often obligatory than the intranidal symbionts and guests. Some of these extranidal relationships are among the most engaging of all, for they undeniably certify that natural selection has incubated some rather circuitous approaches to finding food.

Midden Detritivores, Scavengers, and Predators

Phorid flies are the most numerous denizens of the middens and surely the most common Diptera found with the New and Old World army ants (Kistner 1982). The family Phoridae, sometimes called the humpbacked flies, is a highly variable group whose larvae may feed on decaying animal and plant matter and fungi, and may also be parasitoids of other insects. Although species from all six of the family's subfamilies have been found with social insects, most belong to the subfamily Metopininae (Kistner 1982). The females of most myrmecophilous species are flightless, their wings either absent or reduced to small, setae-studded appendages. The males have large wings and fly (Rettenmeyer and Akre 1968, Kistner 1982). Among the specimens collected with army ants, females are far more abundant than males.

Phorids associate with numerous species of *Eciton, Labidus, Neivamyrmex,* and *Nomamyrmex,* but few if any are host specific (Rettenmeyer and Akre 1968, Disney and Kistner 1989). These ecitophilous phorids can be found running at the beginning and end of emigration columns. They are the most abundant myrmecophilous insects in the foraging columns, and they can be seen moving among or flying over diurnal swarm raids (Rettenmeyer and Akre 1968). When traveling in the columns, phorids run in a zigzag pattern and deviate beyond the edges of the trail by as much as 2 to 4 cm.

Rather low numbers of phorids have been recovered from bivouacs, and Rettenmeyer and Akre (1968) concluded that the large numbers of phorids they found in the middens of *E. burchelli* and *E. hamatum* supported the hypothesis that most species are scavengers. They feed on dead ants and prey refuse, probably do no harm to their host colonies, and may play a beneficial role by eliminating refuse (Rettenmeyer and Akre 1968). Species that are strictly myrmecophilous and depend on the ants to some extent emigrate with their hosts. R. H. Disney and Kistner (1989) surmised that the myrmecophilous phorids are not socially integrated with their hosts.

Among the phorids associated with the Old World army ants are six species from the subfamily Aenigmatiinae, all belonging to the genus *Aenigmatopoeus* and known only from wingless females (Figure 5.9). Like the ecitophilous metopinines, they can be found in foraging and emigration columns, especially at the ends, and they live in the middens of the driver ants, *Dorylus* (*Anomma*), and *Aenictus* (Kistner 1982). I have seen these dorylophiles in an emigration column of *A. asantei.* They were present in the column only when the workers were carrying brood and were the predominant myrmecophile. They appeared as small, gray hemispheres quickly darting to and fro, sometimes stopping abruptly and reversing direction, like so many billiard balls skittering randomly across a narrow billiard table.

Flies of the family Sphaeroceridae are also found in middens, and in foraging and emigration columns, of New and Old World army ants (Borgmeier 1931, Richards 1968, Kistner 1982). The habits of these puzzling, small black or brown flies are as yet undisclosed, although it appears that the adults, at least, feed on the remains of prey and dead host ants in refuse deposits (Kistner 1982).

Predators also frequent the middens, where they prey either on the army ants or on other visitors. The most common staphylinid beetle found in the refuse deposits of *E. burchelli,* for instance, is just such a predator. As many as a hundred individuals of *Tetradonia marginalis* may be found in a single large deposit. This aggressive beetle is also common in emigration columns (Akre and Rettenmeyer 1966) and is discussed below.

Facultative Predators

Like highwaymen lurking about the byways of eighteenth-century England, predaceous beetles frequent the foraging and emigration trails of army ants and by stealth or brute force steal the ants' prey or brood. Carabid beetles of the genus *Helluomorphoides,* for instance, have been observed absconding with the prey and brood of three species of *Neivamyrmex* (Plate 14B). These beetles, which are much larger than the ants, run with both foraging and emigration columns, in either direction, often straddling the columns, sometimes plowing right through them (Plsek et al. 1969, Topoff 1969). In laboratory experiments, the beetles could detect and follow the chemical trails of the ants. When assaulted by the ants, they repelled their attackers with a defensive chemical spray, a mixture of formic acid and n-nonyl acetate (Eisner et al. 1968, Plsek et al. 1969).

Helluomorphoides is primarily nocturnal and a voracious eater. Topoff (1969) observed that when a beetle contacted a worker carrying brood, the ant usually dropped its burden, which the beetle either ate right there or carried off to a place near the column. In two instances, Topoff (1969) saw beetles forcibly take larvae from *N. nigrescens* workers. In each case the beetle grasped its prey in its mandibles, crooked its head upward, and lifted both ant worker and larva. It held them off the ground for almost two seconds, until the ant released the larva and fell back to the substrate. The beetles feed most intensively on nights when *N. nigrescens* colonies emigrate with their larval brood, and one *H. latitarsis* was seen to eat 28 larvae. Whenever a beetle encountered a brood cache—a temporary stockpiling of brood along the emigration route—it feasted until satiated (Topoff 1969). In fact, the beetles would immoderately devour so many ant larvae that their abdomens swelled and the abdominal sclerites separated like those of an egg-swollen army ant queen. *Helluomorphoides* does not inhabit the army ant nest or bivouac.

The staphylinid *Tetradonia marginalis* is one of the most common ecitophiles in and along the foraging and emigration columns of *Eciton burchelli* and *E. hamatum.* It too does not live within the bivouac but inhabits the area around the nest. This beetle is associated with foraging columns at all times of day but is most numerous at the beginning of emigrations (Akre and Rettenmeyer 1966). It commonly runs in the emigration columns, at the center of the columns when traffic is light and

Figure 5.22. A staphylinid beetle, genus *Tetradonia*, attacking a dying *Nomamyrmex* worker in a laboratory nest. A predator of New World army ants, *Tetradonia* is common in colony refuse deposits and along foraging and emigration columns. (Photograph by Carl W. Rettenmeyer, Connecticut Museum of Natural History.)

at the edge when traffic is heavy. Some individuals fly from one colony to another. *T. marginalis* attacks workers of all sizes, injured and uninjured, at the edge of the columns or at the periphery of the bivouac (Figure 5.22). Most attacks involve grasping the ant's hind leg and pulling the ant backward. Uninjured workers usually escape; those that do not are dragged 50 to 250 cm from their sister workers. In laboratory nests, *T. marginalis* trimmed the legs from the body of its captive and then separated the three major body regions, feeding on the fluids that came from the openings of the disconnected parts (Akre and Rettenmeyer 1966).

Swarm Followers
Like sutlers, the camp followers of military history who pursued armies and peddled provisions to soldiers, a variety of insects and vertebrates follow swarm-raiding army ants on their rough-and-tumble

excursions—battles of sorts—for food. The analogy has its weaknesses, but there can be little doubt that many swarm followers rely on army ants for their living, for they prey on the arthropods flushed by the advancing tide of foraging ants.

Conspicuous among the swarm followers are flies representing three families: Calliphoridae, Conopidae, and Tachinidae. The most ubiquitous geographically are conopids of the genus *Stylogaster*, which are associated with both New and Old World army ants (Figure 5.23). Although the association of these flies with army ant swarms was recorded in the last century (Bates 1863, Townsend 1897), little was known of their biology then. C. H. T. Townsend (1897, p. 23) noted their presence at a swarm of *Eciton:* "The specimens of *Stylogaster* hovered continually over the ants, now and again darting at them, without doubt for the purpose of oviposition in their bodies." He was not completely correct. In the New World, *Stylogaster* is associated with swarms of *Labidus praedator* and *Eciton burchelli.* Like elfin helicopters, these flies continually hover above the ground, concentrated at the front of the swarm or as much as 2 meters beyond it. What Townsend missed was the fact that the flies were darting at fleeing cockroaches rather than at the ants (Rettenmeyer 1961). In addition to parasitizing cockroaches, *Stylogaster* flies deposit their eggs on tachinid flies of the genera *Calodexia* and *Androeuryops* that attend the same swarms (Figure 5.24).

The six species of *Stylogaster* associated with ecitonine swarms probably find the ants olfactorily but locate their hosts visually. They are either unable to find hosts without the ants or much less efficient at doing so in the ants' absence (Rettenmeyer 1961). *Stylogaster* eggs, which are probably laid with a stabbing movement of the abdomen, possess a pointed end bearing one or two pairs of recurrent spines, not unlike a harpoon, that impale the host's cuticle (Figure 5.24) (K. G. V. Smith 1967). Old World *Stylogaster* species commonly parasitize flies of the families Calliphoridae, such as *Bengalia depressa,* which are associated with driver ant swarms, or Muscidae, like the genus *Dichaetomyia,* which are associated with dung (K. G. V. Smith 1969).

The tachinid flies found with the swarm-raiding ecitonines have no known analogues in Africa. Little is understood of the one species, *Androeuryops ecitonis,* recorded at army ant swarms. Males are twice as abundant as females around swarms; the female is oviparous; the eggs must be laid inside the host (based on ovipositor morphology); and the hosts are unknown (Rettenmeyer 1961). As with *Stylogaster, Calodexia* and *Androeuryops* flies are found at the front of the ant swarm, but they do not hover there; instead they rest on low objects near the swarm and frequently move to avoid the advancing ants. Thirteen species of *Cal-*

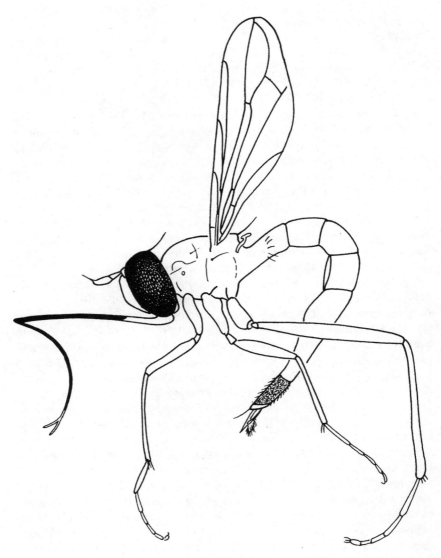

Figure 5.23. An adult female *Stylogaster malgachensis*. (Redrawn, by permission of the Royal Entomological Society, from K. G. V. Smith 1967.)

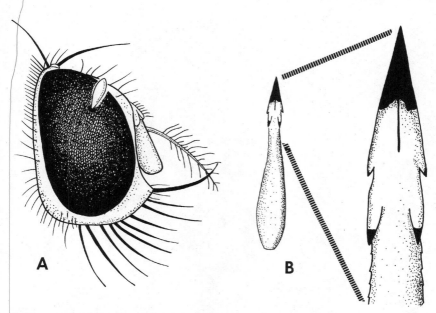

Figure 5.24. A *Stylogaster* egg impaled in the eye of a muscid fly (A), and the egg of *S. westwoodi*, including an enlarged view of its harpoonlike tip (B). (Redrawn, by permission of the Royal Entomological Society, from K. G. V. Smith 1967.)

odexia were collected over swarms on Barro Colorado Island, Panama (Rettenmeyer 1961). Unlike *Androeuryops,* males are rare among the thousand or more females that can be found with large swarms of *E. burchelli. Calodexia* flies probably locate ant swarms by odor and appear at the swarms within 10 minutes of their start, an efficient and expedient achievement. These larviparous flies deposit their larvae on the body surfaces of crickets and cockroaches (Rettenmeyer 1961) (Figure 5.25).

Tropical Old World flies of the calliphorid genus *Bengalia* frequent the territories of various ants, including such genera as *Camponotus* and *Leptogenys,* and rob the workers of prey and brood. They generally grasp the prey or brood with their forelegs (or stab it with their proboscis) and snatch it from the startled worker (Maschwitz and Schonegge 1980). *Bengalia depressa* commonly attends the swarm raids of the African driver ants. W. A. Lamborn (1913–1914, p. cxxvii) depicted the behavior of this fly at a *Dorylus (Anomma) nigricans* swarm:

I then carefully watched a fly hovering over the ant-column. It suddenly swooped down and rose instantly with an ant pupa, with the driver that had been carrying it still hanging on, fixed to its proboscis. The fly carried

Figure 5.25. A fly of the genus *Calodexia*, family Tachinidae, resting on a leaf above a swarm raid of *Eciton burchelli*. (Photograph by Carl W. Rettenmeyer, Connecticut Museum of Natural History.)

this burden for about a foot then dropped it and alighted on the ground near by. The ant started to run away with the pupa, but the fly pursued it, again impaled the pupa and started a tug-of-war with the ant. Neither side had any advantage, and then the fly rose again about three feet into the air with the pupa and ant and after a flight of about eighteen inches let them fall. The ant being discomposed by this procedure let go of the pupa, and no sooner had it done so than the fly seized it and, flying off with it triumphantly, settled near by.

Bengalia is conspicuously different from other calliphorids in its possession of a rigid proboscis endowed at its apex with teeth. It is well adapted for piercing and stealing prey and brood and for sucking hemolymph from these pillaged spoils (Bequaert 1922).

Numerous species of flies attracted to army ants employ devious strategies for infiltrating the colonies or middens with their eggs or larvae. Rettenmeyer (1961), for instance, many times saw a species of Muscidae lay its eggs near the bivouac of *E. burchelli*. The eggs were picked up by the workers and carried into the bivouac or middens. Subsequently, Rettenmeyer (1961) was able to rear a large series of muscids from material found in the middens. Lamborn (1914–1915) observed at least two species of Calliphoridae and one of Anthomyiidae ovipositing in or near the subterranean nest of *D. (A.) nigricans* in southern Nigeria.

Calliphorids of the genus *Tricyclea (Zonochroa)* dropped eggs into the funnel-like openings of the nest or on the soil surrounding the entrances. The latter eggs were eventually covered with soil by the workers as they went about their excavations. Another calliphorid oviposited directly into the soft soil of the nest surface, and the one anthomyiid "hovered over one particular opening made by the ants in the ground and then let drop a number of eggs—as many as six—in rapid succession" (Lamborn 1914–1915, p. vii).

In East Africa, W. H. Thorpe (1942) was surprised to see the muscid *Stomoxys ochrosoma* hover about an inch above a foraging column of *D. (A.) molestus* and drop its large egg or larva in front of a worker returning without prey. The worker immediately picked up the egg and carried it toward the nest. *Stomoxys* adults, like the common New World stable fly, *S. calcitrans*, feed on the blood of mammals by biting them with their piercing-sucking mouthparts. "From the rather variable scavenging habits of *calcitrans* in the tropics," Thorpe (1942, p. 40) conjectured, ". . . the larva of *ochrosoma* is more likely to be a scavenger in the bivouacs of the army ants than a predator on the ants themselves; but even so there must be some very delicate quality possessed by the egg or larva, which prevents it being devoured or injured by the worker ants without rendering it too distasteful to be carried by the foragers."

When observing army ants in Brazil 160 years ago, M. Lund (1831, p. 120) recorded that troops of army ants were constantly followed by a flock of birds, and that one bird species "announce au loin par son cri monotone et lugubre la presence de ces troupes" (announces from afar, with its monotonous and woeful cry, the presence of these troops). Henry Walter Bates (1863, p. 371), who also watched army ants in Brazil, noted that "when the pedestrian falls in with a train of these ants, the first signal given him is a twittering and restless movement of small flocks of plain-coloured birds (ant-thrushes) in the jungle." Thus, the calls of swarm-following birds have long been known to mark the presence of army ants. As R. A. Johnson (1954, p. 41) explained:

On a walk through the forest in tropical America, long periods may pass without the glimpse of a bird. Then, suddenly, all about one hears the chirring, twittering and piping of birds, and sometimes a dim murmur, as if a light shower were striking the leaves of the forest floor. This gentle pattering—it soon becomes clear—is caused by the frantic fluttering and hopping of countless insects trying to escape a swarm of raiding army ants. The insects and other arthropods driven from cover by the army ants provide a readily accessible food supply for the birds. The excited voices of those first finding the ant raid attract other birds to the scene. Before long, a group of many individuals and several species is accompanying the ant swarm.

Nineteenth-century naturalist Thomas Belt (1874, p. 19) correctly assessed the situation in Nicaragua when he noted that "several species of ant-thrushes always accompany the army ants in the forest. They do not, however, feed on the ants, but on the insects they distrub." And explorer Paul Du Chaillu (1861, p. 319) made parallel observations a decade earlier in central Africa while watching the bird *Alethe castanea:* "They fly in a small flock, and follow industriously the bashikouay [driver] ants in their marches about the country. The bird is insectivorous; and when the bashikouay army routs before it the frightened grasshoppers and beetles, the bird, like a regular camp-follower, pounces on the prey and carries it off. I think it does not eat the bashikouay."

Raconteur adventurers, naturalists, and field biologists have sometimes mistakenly assumed that these birds indiscriminately eat flushed prey and army ants alike. Although E. O. Willis and Y. Oniki (1978) found heads of army ants among the milled contents of bird stomachs, they suspected that the ants were inadvertently eaten because they had been attached to the real prey. Ingestion of army ants by swarm-following birds appears more accidental than intentional. Besides, the fact that some myrmecophiles mimic army ant morphology suggests that visually cued predators such as birds select for mimicry—to wit, looking like an army ant is a good way to avoid becoming a hungry bird's lunch (Willis and Oniki 1978). Only once has an army ant follower—albeit an infrequent follower—been observed to prey on the army ants themselves. D. F. Stotz (1992) saw a buff-throated saltator deliberately gleaning a tree trunk of foraging *Eciton burchelli* workers, picking off the ants one by one as they ran by. Birds are attracted to ant brood, though, and Willis and Oniki (1978) speculated that army ants emigrate at night to avoid the loss of brood to birds. Driver ants, however, have their own faithful contingent of avian followers and emigrate at any time of day or night. It is questionable whether bird predation of ant brood is detrimental enough to select for nocturnal emigrations.

Ant following by birds is essentially restricted to swarm raiders, which predictably flush prey in advance of their frenzied inundation of the forest floor. In Africa, avian swarm following is probably limited to the driver ants *Dorylus* (*Anomma*) *molestus, D.* (*A.*) *nigricans,* and *D.* (*A.*) *wilverthi.* Birds have not been reported at the epigaeic raids of Asian *Aenictus* (Willis and Oniki 1978). Although *Eciton rapax* in the New World attracts a few birds to its unspectacular raids, two other ecitonine species, *Labidus praedator* and *E. burchelli,* account for most bird records. Because *L. praedator,* which forages much more frequently in the rainy season than in the dry season, does not conduct day-long raids and fails to forage every day, it is not a dependable source of flushed prey for the birds and is often ignored by them (Rettenmeyer 1963b, Willis and Oniki 1978). *E. burchelli,* on the other hand, is a dependable source of food, as it forages nearly every day in both the nomadic and statary phases of its functional reproductive cycle. During the nomadic phase, its foraging expeditions are all-day events. Willis and Oniki (1978, p. 245) discussed this predictability: "Regularity in swarming, with three or so out-of-phase colonies per km² (on Barro Colorado Island in Panama), allows some ant-following birds to be 'professional'—they get over 50 percent of their food, often nearly all of it, by following army ants. They shift to nomadic and large colonies when possible, but seldom are without a colony to follow. Since the ants swarm in dry or wet seasons or weather, there is no break in food availability."

In the Neotropics, about 50 species of birds, belonging to four families, follow army ants regularly. Most are professionals, meaning that they obtain more than 50 percent of their food near the ants (Oniki 1972). The four families are the cuckoos (Cuculidae), woodcreepers (Dendrocolaptidae), antbirds (Formicariidae), and tanagers (Thraupidae), but in none of these families are the regular ant followers more than a small percentage of the total number of species (Plate 17A). Of the 46 families of Neotropical land birds, 25 have at least 1 species recorded to capture army ant–flushed prey (Willis and Oniki 1978).

Several migrant species of North American birds that winter in Central and South America follow swarms of army ants. These facultative swarm followers include Acadian flycatchers, wood thrushes, and Kentucky warblers (the Kentucky warbler can be a regular visitor [Whitacre and Ukrain 1990]). Such migrants are nearly always subordinate to resident birds, which restrict them to the periphery of the swarm. They are, accordingly, more frequent at the smaller, infrequent swarms of *L. prae-*

dator, where competition is less intense (Willis 1966). The position of migrants at the periphery and their penchant for following the unpredictable swarms of *L. praedator* suggest that resident species may exclude them when food is not particularly abundant (Willis 1966).

In addition to capturing flushed arthropods, some swarm-following birds catch *Anolis* and *Sceloporus* lizards and frogs, some of which may themselves be feeding on flushed prey. And to add yet another level to the vertigo-producing heights of this pyramid of predatory relationships, raptorial birds (falcons and hawks) are attracted to the raucous swarm-following flocks and may occasionally attack and eat antbirds (Willis et al. 1983, Coates-Estrada and Estrada 1989, Haemig 1989).

Like *L. praedator*, African driver ants appear to be undependable sources of food for swarm followers (Karr 1976). Driver ants are non-phasic species and may forage irregularly; they nest within the ground; and in dry weather they may remain sequestered in their subterranean nests and tunnels for long periods. The birds that follow driver ants are mostly horizontally perching thrushes (Turdidae) and bulbuls (Pycnonodidae), which differ both behaviorally and taxonomically from the tree-climbing woodcreepers and vertically clinging antbirds of the New World tropics (Willis 1985a). As a consequence of the differences in perching behaviors, the many vertical saplings and trunks common to the forest are left vacant by the driver ant followers. From their horizontal perches, these birds must sally farther to capture prey than birds that drop from vertical perches (Willis 1985a).

Based on his studies on Barro Colorado Island, Johnson (1954) classified the mixed bird aggregations following army ant swarms into feeding aggregations, composed of birds that associate primarily in connection with the army ant swarms; and social aggregations, comprising birds that associate independently of the army ant swarms but do attend them at times. The feeding aggregations include birds that follow the army ants without reference to territory—for instance, the antbirds—and birds such as the motmot that attend the swarms only when they are near or pass through the birds' territories. Similarly, the social aggregations include two types of birds: those, like some woodcreepers, that travel in the forest in mixed social groups and readily follow the ant swarms for long periods, and those, including small antwrens, that travel in mixed social groups and briefly and temporarily join other birds attending ant swarms.

Swarm-following birds in the Neotropics divide food in two ways: by dominance and by perch type. In the former, birds crowd about and over an army ant swarm. The larger birds occupy the central and best zone and chase off intruding medium-sized birds to the next, less desir-

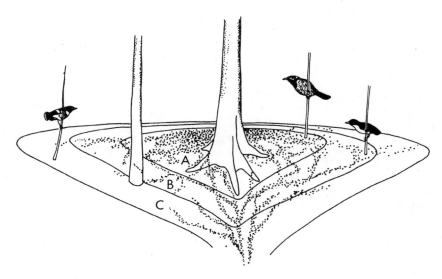

Figure 5.26. Swarm-following birds of the Neotropics crowd about and over army ant swarms; the larger birds occupy the central, and best, zone around a swarm (A) and chase off intruding medium-sized birds to the less desirable concentric zone (B); smaller birds are consigned to the inferior outer zone (C). (Redrawn, by permission of Y. O. Willis and Annual Reviews Inc., from Willis and Oniki 1978, *Annual Review of Ecology and Systematics*, vol. 9, © 1978 by Annual Reviews Inc.)

able, concentric zone; these, in turn, exclude smaller birds, consigning them to the inferior outer zone (Figure 5.26). For example, in Mexico, the central foraging zone near the swarms of both *E. burchelli* and *L. praedator* is occupied predominantly by birds weighing 31 to 41 grams; birds weighing 8 to 20 grams are most commonly found at the periphery at *E. burchelli* swarms and in the middle and peripheral zones at *L. praedator* swarms (Coates-Estrada and Estrada 1989).

Division of niches by perch type is rather more complicated. Ground birds, like ground-cuckoos, although large and potentially dominant, are confined to the periphery of the swarm; climbers, such as woodcreepers, form dominance hierarchies on large perches; and clingers, mostly antbirds, take slender vertical perches and horizontal twigs (Figure 5.27) (Willis and Oniki 1978). Edwin O. Willis has produced an impressively large and detailed body of work on the behavior of swarm-following birds, especially those of the Neotropics, and the interested reader is referred to those contributions (1967, 1972, 1981, 1985a,b,c, 1986a,b,c, and others).

Swarm-following birds have competitors for the bounteous supply of

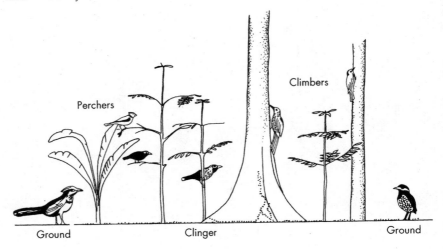

Figure 5.27. Perch type strongly affects the division of niches. Ground birds, like ground-cuckoos, are confined to the periphery of the swarm; climbers, such as woodcreepers, form dominance hierarchies on large perches; and clingers, mostly antbirds, take slender vertical perches and horizontal twigs. Perchers do not usually find places unless antbirds are absent or horizontal perches are abundant. (Redrawn, by permission of Y. O. Willis and Annual Reviews Inc., from Willis and Oniki 1978, *Annual Review of Ecology and Systematics*, vol. 9, © 1978 by Annual Reviews Inc.)

food stirred up by the advancing swarm. Marmosets (genus *Callithrix*, family Callitrichidae) in Brazil opportunely exploit this food source when they encounter swarms of *E. burchelli* and *L. praedator* (Rylands et al. 1989). The marmosets position themselves on perches up to a meter above the ground at the front of the swarm and spend as much as two or three hours grabbing flushed arthropods, principally crickets, grasshoppers, cockroaches, and spiders. With simian audacity, they also snatch prey already captured by the ants. But before eating these purloined morsels, the marmosets shake the prey to dislodge and remove the army ant captors.

During an early Honduran afternoon in 1976, Boyce Drummond beheld a curious sight: flying low over the leading edge of an *E. burchelli* swarm were six butterflies: two males of the genus *Graphium*, family Papillionidae (swallowtails), and females of two species of the genus *Mechanitis*, family Nymphalidae, subfamily Ithomiinae. Flying about 2 feet above the substrate but occasionally dipping down to ground level, these butterflies advanced with the swarm. Drummond (1976) theorized that the butterflies were attracted to the unpleasant feceslike odor that emanates from *E. burchelli* colonies and may contain elements of the male *Mechanitis* sex pheromone (explaining the presence of only *Mechanitis* females). Dissatisfied with Drummond's so-called reproductive odor hypothesis, A. M. Young (1977, p. 190) proposed a feeding hypothesis to account for the presence of the butterflies. He noted that ithomiine butterflies are attracted to fresh bird droppings as nutrient sources, that these butterflies probably orient to such nutrient sites by odors of decay, that such odors trigger food-searching behavior in these butterflies, and that *Graphium* and *Mechanitis* are being fooled by the *E. burchelli* swarm odors.

Indeed, Neotropical butterflies, including ithomiines, do feed on bird droppings (Young 1984), which contain uric acid and partly digested proteins, nitrogenous nutrients necessary for egg production (Ray and Andrews 1980). *Mechanitis* females (Plate 17B) and females of the ithomiine genus *Melinaea* assemble in high densities at *E. burchelli* swarms and feed on bird droppings in the immediate vicinity (Ray and Andrews 1980). The source of the bird droppings: swarm-following birds, especially of the family Formicariidae. T. S. Ray and C. C. Andrews (1980) postulated that army ant swarms provide a predictable source of nutrients in the form of bird droppings, and that swarm odors may permit the butterflies to locate fresh droppings on which to feed. Gerardo Lamas (1983) discovered an additional 15 species associated with army ant swarms, all belonging to the family Hesperiidae, the skippers. Clearly, there is more to be learned about these and other swarm followers, and no doubt the complexity of their biotic interplay will continue to challenge our imaginations.

Predators

Excluding the hurly-burly assemblage of army ant predators that rank as symbionts, guests, or associates, army ants are subject to the gustatory instincts of numerous predators, invertebrate and vertebrate alike, especially when foraging or emigrating aboveground. Although the morphology and behavior of ants are generally viewed as adaptations to resource acquisition strategies, J. H. Hunt (1983) hypothesized that predation, especially vertebrate predation, may be a selective force acting on morphological and behavioral features. Hunt (1983, p. 100) admitted, however, that "trunk trail foraging in these species [of army ants] may be more strongly related to resource acquisition than to predator resistance," and that worker polymorphism seems more related to resource gathering than to predation.

Invertebrates

Not the least important invertebrate predators of army ants are other ants, an ecological irony of sorts, as ant nests are quite often raided and plundered by New and Old World army ants. Foraging or raiding for prey as army ants do, however, is a hazardous profession, and numerous workers return to the nest from such forays injured—limping stragglers like so many human casualties retreating from some horrendous battle. They commonly journey to the nest long after the other workers have completed their return, and vulnerable, they often fall prey to other ants (Gotwald 1982) (Plate 3A). ˙

Ants of the subfamily Myrmicinae, especially of the genus *Cremato-gaster*, are prevalent among these opportunistic predators. On one occasion, in unusual circumstances, I observed hundreds of small *Crematogaster* workers dragging off to their nest injured driver ants. An automobile had run over their foraging column, and in the confusion thus wrought, the *Crematogaster* workers were able to go about their business with impunity. None were attacked by the driver ants (Gotwald 1972b).

Cohic (1948) also recorded an encounter between *Crematogaster* and driver ants, and I have seen foraging *Dorylus* (*Anomma*) *gerstaeckeri* workers attacked by ants of the myrmicine genus *Pheidole* (Plate 3A). In another unusual circumstance, I saw the ponerine ant *Paltothyreus tarsatus* carry off larvae and pupae of the driver ant *D.* (*A.*) *nigricans* that had been discarded in soil removed from a colony's nest as it was being excavated (Gotwald 1972b).

Newly emerged army ant males are probably subject to intense predation from all quarters. At night they are attracted in substantial numbers to incandescent lightbulbs, as are numerous predators that take advantage of the assembling feast. In West Africa, light-attracted males are captured by the ponerine ant *Megaponera foetens,* as well as by spiders (Plates 18A, 19) (Gotwald 1982).

Most fascinating of the formicid predators of driver ants is the African weaver ant, *Oecophylla longinoda.* This arboreal species, of the subfamily Formicinae, is a highly predaceous ant that makes its nests by pulling leaves together and binding them in place with larval silk. Workers "sewing" together leaves carry last-instar larvae about as "living shuttles" which discharge strands of silk from their labial glands (Hölldobler and Wilson 1990, p. 618). One colony may build hundreds of nests distributed over several trees, and each colony defends an "absolute territory," meaning that the territory is defended at all times, regardless of the resources it contains (Hölldobler and Wilson 1990, p. 400). Any insect entering this territory is likely to be captured by these pugnacious but stingless ants, and army ants are no exception.

O. longinoda attacks both emigrating and foraging columns of driver ants, but attacks the columns rather judiciously from but a few vantage points. The assault consists of three phases: (1) the attack phase, in which an individual driver ant is seized in the mandibles of an *O. longinoda* worker, often perched on leaves overhanging the army ant column, and removed from the stream of ants (Plate 18B); (2) the immobilization phase, in which additional *O. longinoda* workers are recruited to spread-eagle the driver ant until it is no longer able to move; and (3) the transfer phase, in which the immobilized driver ant is carried by one or two of the attackers to the *Oecophylla* nest (Gotwald 1972b).

O. longinoda is an effective predator of driver ants for three reasons. First, the weaver ants usually attack the driver ant column at a limited number of points and thus avoid disrupting the movement of the driver ants. Second, they grasp and remove the driver ant worker quickly, thereby avoiding a widespread alarm response. Third, they attack the driver ants from defensively advantageous positions above or at the periphery of the column and thereby reduce their vulnerability to a sudden alarm response (Gotwald 1972b). It is not imprudent to presume that in areas densely populated by *O. longinoda*, where the probability of encountering its territories is high, driver ant colonies may lose several hundred, even thousands, of workers per day to *Oecophylla* predation (Way 1954, Vanderplank 1960, Gotwald 1972b, Dejean 1991). The closely related Asian weaver ant, *O. smaragdina*, the only other extant species in the genus, is also highly territorial (Hölldobler 1983) and commonly preys on *Aenictus* workers whose columns violate its territories (Plate 20A) (Rosciszewsk; and Maschwitz 1994). Captured trespassers are spread-eagled and carried off to the attackers' arboreal nest (W. H. Gotwald, Jr., unpubl. data).

Vertebrates

Vertebrates probably constitute the greatest predatory threat to army ants. Amphibians, for instance, typically feed indiscriminantly on ground-inhabiting ants. Joseph Bequaert (1922) reported that of a total of 1815 ants found in 194 stomachs of five species of Congo region toads, genus *Bufo*, 8 percent were dorylines. There were six species of driver ants represented in the stomach contents of four of the toad species. Three species of African frogs likewise eat driver ants, as do reptiles. Forest skinks of the African genus *Mabuya* apparently even follow columns of driver ants. Y. Sjoestedt (quoted by Bequaert 1922, p. 297) observed *Mabuya raddoni* in Cameroon:

This lizard is one of the most diligent persecutors of driver ants, and wherever one of their columns was seen on the move in or at the margin of the forest, especially after the ants had scattered in search of food, one could be sure to find one or more of these graceful animals preparing for an excellent catch. It was a delight to observe how adroitly the agile lizards would plunge into the crawling swarm, fill their mouth with ants and then retire to a place of safety to devour their booty. Busily engaged in their hunt, they would fearlessly run about the motionless observer and not even hesitate to climb his legs, always twinkling their lively little eyes, on the lookout for possible danger.

Insectivorous birds most definitely include army ants in their diets. R. S. Baldridge et al. (1980) speculated that New World army ant males are taken by such nocturnal predators as owls, nightjars, and nighthawks. Under experimental conditions, live *Neivamyrmex* males were eaten by mockingbirds, cardinals, lark sparrows, tufted titmice, and meadowlarks. In India, *Dorylus* is eaten by birds (Bequaert 1922); and in Africa, driver ants are devoured by several avian species, including the guinea fowl, *Phasidus niger* (Chapin 1932). Even the domestic chicken takes its share of the army ants that forage in village refuse heaps. Undaunted by the aggressiveness of the disturbed ants, the chicken, steps gingerly among the foragers, picking up and swallowing individual workers, all the while avoiding any other contact with the ants (Gotwald 1972b).

Among mammalian predators of army ants, pangolins (genus *Manis*), the scaly anteaters of the Old World tropics, probably take large numbers of dorylines. These curious mammals are at once recognizable by their covering of overlapping scales, short legs that end in curved claws, and the long, weighty tail that counterbalances the rest of the body, making a teeter-totter of the whole animal when it rears up slightly on its hind legs. The pangolin's stomach, endowed with horny teeth, is specialized as an organ of trituration well adapted to its diet of whole insects (Bequaert 1922). Driver ants are common in the stomach contents and feces of pangolins. The stomach of one *Manis gigantea* contained 6 liters of termites in its anterior portion and an equal amount of driver ants in the posterior portion (Bequaert 1922). Herbert Lang (in Bequaert 1922, p. 323) described the behavior of one pangolin as it dined on a column of driver ants:

Sitting on its hind limbs and with its tail steadying its movements, the fore part of the body was swung about freely. The claws of the fore limbs were kept busy removing those of the fierce assailants that, in spite of the oft repeated shivering movements of the scaly armor, succeeded in gaining a hold. Lashing its sticky tongue through the confused crowds, the ant-eater lost no time in moving back and forth along the ant column as quickly as the dense clusters vanished into its mouth. Its hunger satisfied, it at once retreated, freeing itself of the few army ants that had managed to dig their mandibles into the soft parts of its hide.

Driver ants have also been recovered in the stomach contents of the African mongoose, *Bdeogale nigripes* (Bequaert 1922). The stomach of a female aardvark examined by S. Patrizi (1946) contained in excess of 100,000 *Dorylus* (*D.*) *helvolus*, in all stages of development, suggesting that the aardvark had pillaged a *Dorylus* nest. A subterranean but unidentified species of *Dorylus* was found in fecal samples of the aardwolf, *Proteles cristatus*, which is normally a termite feeder. Indeed, H. Kruuk and W. A. Sands (1972) observed that certain prey species usually not present in the feces, including *Dorylus*, suddenly became abundant in some samples collected during the rainy season.

Possibly the most intriguing army ant predators, at least from the standpoint of the predator's behavior, are mountain gorillas and chimpanzees. The large size of the mountain gorilla, *Gorilla gorilla beringei*, makes it impractical for this primate to rely on sparsely distributed invertebrates as a protein source. Nevertheless, gorillas do intentionally eat some invertebrates and inadvertently ingest many others, especially insects, along with the vegetation they eat. These invertebrates may supply the gorillas with vitamin B_{12} and other required micronutrients not available in plants (Harcourt and Harcourt 1984).

In the Virunga Volcanoes region of Rwanda and Zaire, mountain gorillas have been observed feeding on driver ants (Watts 1989). Eleven ant-eating episodes were recorded, six in their entirety. Although the gorillas sometimes discovered the driver ant nests while engaged in other activities, some individuals recognized the undisturbed nests, even though few or no ants were present, and excavated the nests. The process of nest excavation was described by D. P. Watts (1989, p. 122):

> To obtain ants, gorillas rush up to a swarm or reach into a nest and grab a handful, then retreat hastily and frantically eat them from their hand. Their dense hair offers some protection from the ants' fierce bites, but ants that burrow through to the skin cause obvious discomfort. After eating a handful, a gorilla picks individual ants out of its hair with its fingers or lips and eats them and vigorously brushes others onto the ground. Often the gorilla then dashes back to the nest to repeat this sequence. As nest excavation proceeds, individuals thrust their arms into the enlarged opening to get ants. In five of eight observed sessions, the gorillas eventually reached the eggs and pupae, which they ate eagerly.

The ant-feeding episodes lasted 8 to 35 minutes, and the number of participants ranged from 2 to 12. The average individual ate approximately 3 to 10 grams of ants per session. Watts (1989) concluded that although some individuals avidly consumed driver ants, ant-eating bouts were so rare as to make ants nutritionally unimportant to mountain gorillas. Even though ant eating occurred more frequently among younger individuals, Watts (1989) presumed that it most likely is a result of individual differences in taste and learning experiences rather than differences in nutritional strategies associated with age or sex classes. Indeed, a female that transferred from a group in which ants were not eaten to a group that did eat ants acquired a taste for driver ants and learned to participate with her new group in ant-eating episodes.

Jane Goodall (1963), famous worldwide for her research on chimpanzees at Gombe National Park in western Tanzania, was the first to note that the East African chimpanzee, *Pan troglodytes*, not only eats driver ants but collects them by employing modified sticks—a clear example of tool use. W. C. McGrew (1974, p. 501), who called this ant-harvesting behavior ant dipping, observed that "the free-living chimpanzee is now well established as the paramount nonhuman tool maker and user." The Gombe chimpanzees frequently eat driver ants, *Dorylus* (*Anomma*) *molestus*, in January and occasionally do so during September through November and February through May (van Lawick–Goodall 1968). Apparently, chimpanzees locate driver ants visually, either by finding a column of the ants moving across an open space (not unlike the method the stalwart biologist uses, but the comparison shall rest here), by seeing and recognizing the cratered surfaces of nests, or by discovering previously exploited nests, which are obvious for their abandoned dipping tools and the crushed vegetation that surrounds them.

McGrew (1974) noted that the majority of ant-dipping episodes involved nests that had been used by chimpanzees at least once before. Although the discovery of some previously used nests was clearly accidental, in other instances the chimpanzees knew the location of the nest from prior experience. One individual made a dipping tool and then walked without hesitation to an active driver ant nest 75 meters away.

Figure 5.28. A chimpanzee feeding on driver ants harvested by ant dipping. A. First, the chimpanzee inserts an ant-dipping tool, or wand, into a nest of *Dorylus* (*Anomma*) *molestus*. B. As the disturbed driver ant workers climb the dipping tool, the chimpanzee monitors their progress. C. When the ants have climbed about three quarters of the way up the wand, the stick is withdrawn from the nest; the left hand then slides up the tool in a pull-through, catching the ants in a mass. The chimpanzee's mouth is open, ready to receive the ants. D. The ants are chewed and ingested at the end of the pull-through. (Photographs by W. C. McGrew. Reprinted, by permission of W. C. McGrew and Academic Press, from McGrew 1974.)

When a chimpanzee finds a column of ants, it immediately begins making a dipping tool. When it locates a previously exploited nest, the chimpanzee reactivates it by poking into it with tools left behind by former visitors. When exploiting a newly discovered nest, however, the chimpanzee rakes the surface of the nest with its fingers, creating a hole up to 15 cm wide and deep. Then, "the chimpanzee breaks off one or more branches of living woody vegetation, usually from the nearest appropriately shaped shrub or sapling nearby. The potential tool must be straight, sturdy, long, and without side branches. The chimpanzee then pulls off by hand all or most of the leaves or leaflets. In some cases (with rough-barked plant species?) it may further strip off bark by seizing bits at one end of the branch between the lips and pulling longitudinally" (McGrew 1974, p. 504). The resulting smooth wand averages 66 cm in length and about 1 cm in diameter (McGrew 1974, Sugiyama et al. 1988).

In the dipping sequence, the wand is often held in the favored hand and inserted, slender end first, into the hole created by the earlier raking behavior (Figure 5.28). When the disturbed ants have climbed about three quarters of the way up the wand, the chimpanzee withdraws it and shifts it to a vertical position with the distal end of the stick positioned just below its mouth. It now slides its free hand up the length of the wand in a quick continuous motion (McGrew 1974). This forces the ants into a mass of about 300 ants (McGrew 1974). As its hand moves up the wand, the chimpanzee opens its mouth, and the sliding hand transfers the mass of ants into its mouth. Presumably the ants are so jumbled and disoriented that few bite the chimpanzee before being consumed. McGrew (1974, p. 505) observed that "the chimpanzee's mouth closes and the jaws gnash frantically and exaggeratedly, audibly crushing the ants between the teeth; the ants are then normally chewed and swallowed." The sequence from withdrawal to ingestion takes about two to three seconds.

The average ant-dipping episode observed by McGrew (1974) lasted about 15 minutes, and the likely intake of ants per chimpanzee was about 17.6 grams. He concluded that ants may represent an important food for chimpanzees, especially females, and that the tool technique used by the chimpanzees is complex and sophisticated.

Although some ants are taken as human food—*Oecophylla smaragdina*, the green tree ant, for instance, is made into a paste and eaten as a condiment with curry in Malaysia and Thailand—army ants do not occur on humankind's bill of fare (Gotwald 1986). Just as well! It would seem that army ants have more than their share of insatiable predators, both intranidal and extranidal. They live in a biological community of entangled interspecific relationships where they, the matchless hunters, are often the hunted.

6

The Role of Army Ants
in Tropical Ecosystems

Like the musical instruments of an orchestra that harmonize in the performance of a complex composition, a multitude of living organisms, as various as the tonalities produced by such instruments, interact to form an ecosystem, a unit defined as a biological community and its physical environment. Although at first glance the biotic and physical makeup of an ecosystem may seem random, a discernible orderliness characterizes the evolved interdependencies of its constituents.

An ecosystem is more than the sum of its parts; it is, rather, the sum of its parts and their interactions. This is especially relevant with regard to tropical ecosystems, particularly tropical moist forests, which are notable for their biotic diversity. Army ants figure importantly in tropical and subtropical ecosystems. Their impact on prey populations is considerable, and, as predators of consequence, they even play a role in maintaining species diversity. Because army ants are such enterprising and edacious foragers, and because they share space with human beings, they also affect human economies, in both positive and negative ways. How significant are army ants in the daily cadence of tropical life?

Population Ecology of Army Ants

Ultimately, the dispersal and distribution of ant colonies in any habitat are determined by the aversive interaction of the colonies, conspecific or otherwise. Social insect colonies belonging to the same or related species are generally overdispersed, meaning that the distances between colonies are too uniform to have been arrived at randomly (Hölldobler and Wilson 1990). S. C. Levings and Nigel Franks (1982), for instance, demonstrated the existence of both intra- and interspecific overdispersion among 15 ground ant species on Barro Colorado Island, Panama, and concluded that this spatial pattern was at least partly the result of competition between the colonies. The most negative interactions in this competition, however, occurred between conspecific colonies. Levings and Franks proposed that nest movement and the preferential predation on conspecific queens and incipient colonies by established colonies produces this pattern. Predation on conspecific queens is likely to be adaptive, they suggested, because "the worst kind of neighbor for an established colony is one that has the same foraging habits and food preferences" (Levings and Franks 1982, p. 343).

Although little is known about the dispersion of army ant colonies, especially their interspecific distribution patterns, field research has shown that army ant populations are quite stable. *Eciton burchelli*, for example, has maintained a relatively constant number of colonies on Barro Colorado Island for at least 40 years (Franks 1982a). Using a new censusing technique, Franks (1982a) estimated that there were 55 colonies on the island, or 1 colony per 30 hectares, a number consistent with earlier estimates (Schneirla 1956, Willis 1967).

The reasons for this stability are unclear. Although *E. burchelli* is capable of rapid population growth, such growth has not occurred on Barro Colorado Island. According to Franks (1982c), army ant populations are basically unaffected by competition for cockroaches, crickets, foraging ants, and other transient prey. Lizards, toads, frogs, and other insectivorous species exploit these prey species as well. On the other hand, Franks (1982a,c) proposed that the growth of *E. burchelli* populations may be stabilized by their relation with prey ant populations. Social insect populations may be relatively more constant in size than nonsocial insect populations (Franks 1982a), and although prey ant populations recover more slowly than nonsocial insects from army ant raids, the raids apparently facilitate the founding of new prey ant colonies (Franks 1982c).

Even less is known about the dispersion of Old World army ant colonies. Jean Marie Leroux (1977, 1982), an indefatigable pursuer of army ants, calculated colony densities for the driver ant *Dorylus* (*Anomma*) *nigricans* in Ivory Coast. In savanna habitat, he found one colony per 13 hectares; in forest, four colonies per 13 hectares. These colony densities translate into a driver ant biomass of 88 kg per 100 hectares of savanna and 350 kg per 100 hectares of forest (Leroux

1982). Leroux did not directly address the issue of population stability, although he did document the demise of colonies through death of the queens. Queen death could be attributed to no apparent cause in some instances, and to accidents or to attacks by predators in others (Leroux 1979, 1982).

In a study of the driver ant D. (A.) *molestus* in Kenya, my colleague G. R. Cunningham–van Someren and I twice recorded four colonies nesting simultaneously within a 5-hectare study area (Gotwald and Cunningham–van Someren 1990). This, of course, was a much greater density than that found by Leroux (1977, 1982) for D. (A.) *nigricans*. The habitat of our study area, although internally diverse, was highly modified and may account for the difference in colony densities (see Figure 4.14). Straight-line distances between the neighboring nests ranged from 61 meters to 254 meters. This four-colony overlap lasted only 2 days in the first instance and 24 days in the second. Colony KC-506, which we observed for 432 consecutive days during this investigation, was involved in five collisions with two other colonies, a much higher frequency than the once per colony every 250 days estimated for E. *burchelli* on Barro Colorado by Franks and Fletcher (1983). Given the enormous size of driver ant colonies and swarms, their unspecialized diet, and their enumerated colony densities, it is intuitively difficult to understand the calculated colony density for E. *burchelli* (one colony per 30 hectares) on Barro Colorado Island. Perhaps this low density is a peculiarity related to as yet unspecified restrictions imposed on island populations.

The nature of interspecific competition for similar food resources among army ants and the possibility of interspecific overdispersion as a result remain purely a matter of conjecture. There is, however, evidence of resource partitioning to avoid such competition. A study of 11 sympatric species of Asian *Aenictus*, all of which specialize in ant prey, revealed such a pattern. That is, the *Aenictus* species reduced competition for the same resources by differentially preferring specific taxa (e.g., A. *camposi* preferred the formicine *Paratrechina*), by foraging in different strata (e.g., litter vs. tree trunks), or by favoring a particular prey size (Rosciszewski and Maschwitz 1994). The existence of hypogaeic species confounds attempts to empirically determine the population ecology of congeneric and intergeneric species in shared habitats. Although subterranean species tend to be trophic specialists, their dietary preferences no doubt overlap with those of other specialists and with the less discriminating trophic generalists. For the most part, the population ecology of army ants remains terra incognita.

Impact on Prey Populations

Effect on Prey Abundance

In a study of the effect of foraging by *Eciton burchelli* on leaf-litter arthropods in Costa Rica, G. W. Otis et al. (1986) concluded that organisms that find themselves in the path of an army ant swarm experience one of three possible fates: (1) direct predation by army ants, (2) subsequent predation or parasitization by swarm followers, or (3) escape. Approximately 80 percent of the organisms at the front of the *E. burchelli* swarm escaped, although not all taxa did so with equal success. While *E. burchelli* did not capture spiders, beetles, or flies, 42 percent of the crickets and 36 percent of the cockroach nymphs flushed by the advancing swarm were caught. Invaded ant nests suffered heavy losses of larvae and pupae. Many of the arthropods that escaped the army ants merely postponed death, for they were subsequently killed by ant-following birds or parasitized by the flies that follow the swarm (see Chapter 5) (Figure 6.1).

The same researchers (Otis et al. 1986) found that more arthropods were present in research plots before army ants arrived than remained after they had foraged—hardly a surprising result (Figure 6.2). On Barro Colorado, *E. burchelli* lowered the abundance of cockroaches, crickets, and other prey species by 50 percent, but these prey populations recovered to their preraid levels within a week (Franks 1982c). In an ecological study of the floor fauna of the Panama rain forest, E. C. Williams (1941) concluded, having observed *Labidus coecus* foraging, that "army ants constitute the only predator factor which could be correlated with population densities. The raiding columns of these ants pass through a given area and capture or drive out a large portion of the animal life. In two cases there was definite evidence that army ants had raided the area the day before collection."

Franks (1982c) calculated that a colony of *E. burchelli* on Barro Colorado consumed less than 50 grams dry weight of insects per day, 54 percent of which were ant adults and brood, while ant-following birds

---

Figure 6.1. The potential fates of leaf-litter arthropods in the path of an *Eciton burchelli* swarm. Survival is affected by the probability that army ants will forage in that location and by predation by army ants and swarm-following birds and parasitism by flies. Marmosets and other insectivorous animals attracted to the swarm may also affect survival. The size of each cell in the diagram indicates the relative number of arthropods that encounter that particular fate. (Redrawn, by permission of G. W. Otis and the Association for Tropical Biology Inc., from Otis et al. 1986.)

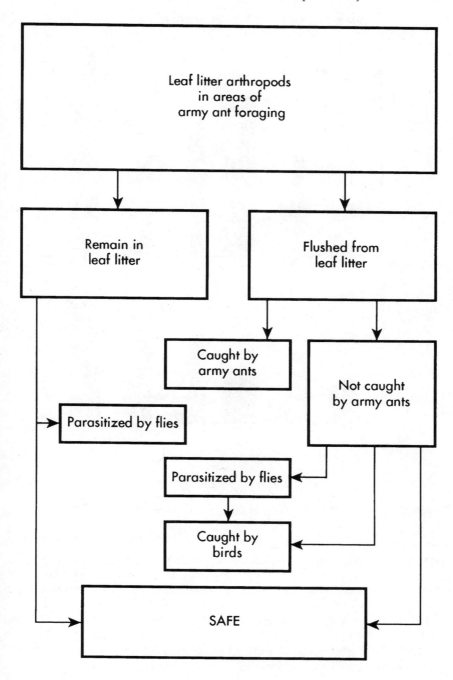

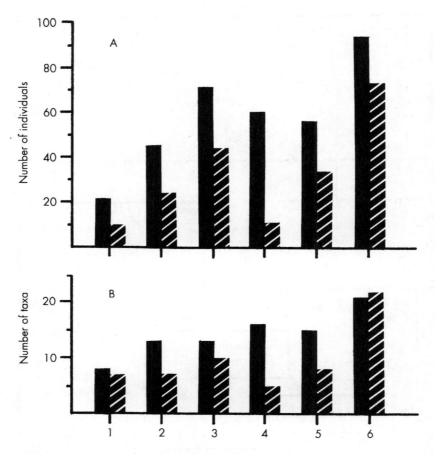

Figure 6.2. Arthropod populations (A) and diversity (B) from six matched pairs of 1-m² leaf litter samples taken before (solid bars) and after (striped bars) the passage of a foraging swarm of *Eciton burchelli*. (Redrawn, by permission of G. W. Otis and the Association for Tropical Biology Inc., from Otis et al. 1986.)

ate roughly 50 grams dry weight of insects—not including ants—flushed by the foraging workers and therefore have a far greater impact on these non-ant arthropods than does *E. burchelli* itself. Thus, a colony of *E. burchelli* and its attendant birds eat about 100 grams dry weight of insects each day. Since there is one colony, and consequently one swarm, of *E. burchelli* per 30 hectares, the colony and its attendant birds eat only 3 grams dry weight of insects, not including ants, per hectare per day. But the ants and their avian associates are not the only insectivores present.

Franks (1982c) wanted to know how the impact of *E. burchelli* and swarm-following birds compared with that of other forest floor insectivores, and how long it took prey populations to recover from a raid.

There are two noteworthy forest floor insectivores on Barro Colorado: the lizard *Anolis limifrons* and the coatimundi *Nasua narica*. Based on previously published research, Franks reckoned that the aggregate feeding rate for the lizards and coatis was 32 to 45 grams dry weight of litter animals per hectare per day—an impressive amount indeed compared with the paltry 3 grams dry weight of insects per hectare per day devoured by *E. burchelli* and its attendant birds. *E. burchelli*, Franks (1982c) concluded, eats much less than its chief competitors.

The impact of army ant predation on social insect prey is more easily assessed than the impact on nonsocial arthropods. Social insects, after all, live in discrete sedentary nests that recover slowly from such attacks. One of the more common and conspicuous ant prey of *E. burchelli* on Barro Colorado, for instance, took 100 days to recover to half of its original population level (Franks 1982c). And the effect of *E. burchelli* on the overall abundance of its ant prey is significant. For example, colonies of its prey species are more than twice as abundant in areas of forest where *E. burchelli* does not occur than where it does nest and forage (Franks 1982b).

Unlike the generalist *E. burchelli*, *E. hamatum* feeds almost exclusively on a single prey, ant brood, and probably has a considerable effect on prey populations (Rettenmeyer et al. 1980). Occasionally, *E. hamatum* preys on social wasps of the subfamily Polistinae (family Vespidae), and although these wasps constitute only 0.5 percent of the prey pieces returned to the bivouac, by dry weight they make up 2 to 10 percent of the total prey. Noting that colonies of *E. hamatum* average 150,000 adults and 60,000 larvae per colony, Rettenmeyer et al. (1980) calculated that the average strong raid brings in 15,000 to 40,000 prey per day and that maximum raids of *E. hamatum* yield as many 90,000 prey, including many minute ant eggs. These figures translate into a daily biomass of between 5.4 and 19 grams dry weight in the former instance and 32.2 grams or more in the latter. These daily foraging yields are considerably lower than those of *E. burchelli* and most likely result from the facts that *E. hamatum* is a predatory specialist and *E. burchelli* a generalist, *E. hamatum* is a column raider and *E. burchelli* a swarm raider, and *E. hamatum* (which has smaller colonies) fields far fewer foragers per raid than *E. burchelli*. (The average *E. burchelli* swarm is 6 meters wide and contains up to 200,000 workers [Franks and Bossert 1983].)

Rettenmeyer et al. (1980) suggested that because army ants tend to crop ant and wasp colonies by taking brood but seldom adults, they are ecologically analogous to grazing animals. After a raid, the surviving adults of the prey colony resume the production of brood, which may be taken by yet another foraging army ant colony. Thus, like the great herbivores of the Serengeti Plain of East Africa, which engage in massive seasonal migrations in pursuit of regenerated grasses (Bell 1971), army ants on the move graze on renewable nutrient resources in the form of social insect colonies. Rettenmeyer et al. (1983) further concluded that predation by army ants must prolong the length of time necessary for ant and wasp colonies to become large enough to produce queens and males.

The effect of army ants on wasp abundance may be profound. In Costa Rica, Young (1979) observed an attack by *E. burchelli* on nests of the vespid *Polistes erythrocephalus*. Seventeen of 34 active nests, arranged linearly on the eaves of a building, were attacked, resulting in 100 percent mortality of pupae and large larvae. Adult wasps fled from both attacked and unattacked nests, many were permanently dispersed as a result. Subsequent observations demonstrated that the *Eciton* attack caused a notable reduction in the building's wasp population.

Commenting on the diet of Neotropical army ants, Rettenmeyer et al. (1983) noted that although termites are abundant in tropical forests, no common species of army ant forages for them. Their observations led them to suggest that *Eciton*, at least, is repelled by the termites' chemical defenses.

In Africa, on the contrary, army ants are well-documented predators of termites. The mound-building termite *Macrotermes bellicosus*, for instance, is commonly attacked by *Dorylus (Typhlopone) dentifrons*, with devastating results (Bodot 1961). Army ant attacks on this species apparently led to its decline, especially on land cleared for planting commercial crops, where it virtually disappeared (Bodot 1967). *Macrotermes michaelseni*, an East African mound builder, is similarly attacked by *Dorylus* (Abe and Darlington 1985). *Dorylus* is a particular threat in later stages of *M. michaelseni* colony growth, especially when reproductives are present. In this late stage of development, only *Dorylus*, of all the ants that attack this termite, can be destructive, because *Dorylus* has a special talent for using the long subterranean termite galleries as access routes to invade the nest and kill the termite reproductives (Abe and Darlington 1985). The same researchers concluded, however, that the distribution of termite colonies, or mound density, is not regulated by ant predation but rather by territory based on the radially arranged subterranean galleries that extend from the nest.

Prey Defenses

The evolutionary refinement of prey detection and capture mechanisms among predators has resulted in a concomitant elaboration of progressively more sophisticated means of defense and escape in prey species. It is not difficult to envision predators as agents of selection in the evolution of prey species. Although successful defense often precedes escape from a predator, some organisms appear to emphasize escape alone, neglecting preliminary defensive systems. Defense and escape mechanisms among organisms attacked by army ants involve predictable behavior patterns such as running or flying, the use of chemical repellents or defensive secretions, and, among social insects, nesting associations with other animals or plants that provide some form of protection and nest characteristics that hamper attacks.

If any escape mechanisms can be described as generalized behavior patterns preadapted to army ant attack—that is, without specialized features selected for by army ant predation—then running, jumping, hiding, climbing vegetation, and flying certainly qualify. Most potential army ant prey escape in these ways (Rettenmeyer 1963b, Schneirla 1971). But some arthropods take a less dramatic approach. These remain motionless when surrounded by raiding workers, stoic witnesses to the hungry tempest, and thereby avoid attack, as if by adaptive aloofness. Walking sticks (orthopterans of the family Phasmatidae) behave in this way (Otis et al. 1986), as do daddy longlegs, or harvestmen (arachnids of the order Phalangida). The long, stiltlike legs of the latter are too thin for the attacking ants to climb (when under attack, an individual phalangid may stand on fewer and fewer of its eight legs until finally, in the interest of continuing equilibrium, it runs off). Heavily armored beetles, especially of the family Scarabeidae, also remain motionless when caught in the midst of army ant raids (Rettenmeyer et al. 1980).

Among the running, jumping, and hiding arthropods are spiders and cockroaches; pillbugs (terrestrial crustaceans) run; crickets jump; and adult cockroaches fly (Otis et al. 1986). In West Africa, large earthworms swiftly slither to escape in advance of foraging swarms; some even scale small plants, a behavior that defies our expectations for apodous, subterranean animals (W. H. Gotwald, Jr., unpubl. data) (Plate 20B). Both earthworms and ants are blind, yet each appears acutely aware of the presence of the other.

Another method of escape effectively separates prey from predator by means of a bridge—one too slender to be negotiated by plant-climbing foragers. Spiders, for example, acrobatically drop from vegetation and hang suspended on silken threads (Swynnerton 1915, Rettenmeyer 1963b). Silk-spinning caterpillars do the same. While observing one driver ant foraging episode in Kenya, I saw six caterpillars elude capture by dropping on threads of silk from the leaves of low vegetation (Gotwald 1972a).

An analogous behavior occurs in slugs (pulmonate terrestrial mollusks) (Gotwald 1972a). In Ghana, I watched workers from a swarm of the driver ant *Dorylus* (*Anomma*) *nigricans* climb a blade of grass about 1.5 meters tall and accumulate in a cluster behind a slug (8 cm long) at the blade's tip. A number of workers attempted to bite the slug but apparently were prevented from doing so by its coat of integumental slime, a feature that also serves as a defensive mechanism against other arthropod predators (Eisner 1971). An increasing number of workers became ensnared in the slime, and eventually the leaf tip bent under the combined weight of the slug and the ants. The slug slid from the leaf and hung suspended from it posteriorly by a thread of slime. There it remained, pendulant, for more than 10 minutes as the thread of slime stretched to a length of 20 cm, at which time it broke. All this time slime-mired ants were trying unsuccessfully to climb down the cord. The thickly matted ground cover obscured the fate of the slug, although the foraging front of the driver ant swarm had by that time moved on.

Although arthropods commonly defend themselves with chemical repellents, few such reactions to army ants have been recorded. G. D. Hale Carpenter (1920, p. 280), of the colonial Uganda Medical Service, reported that "a medium sized Plant bug (Hemiptera) was once seen on a leaf of a bush over which *Dorylus* was swarming, and I was much interested to note that although ants would frequently seize hold of a limb or antenna of the bug they always let go again, and no harm was done to the bug. It is possible that its powerful odour may have been disagreeable to the ants." I often observed pentatomid bugs in West Africa attempting to escape foraging driver ants by climbing to the tops of grasses. As they climbed, they released a chemical, presumably defensive, that was so odoriferous that I occasionally located army ants by smelling the excited stinkbugs. The chemical, however, proved useless against the driver ants (Plate 9).

Another defensive response to *Dorylus* attack observed by Carpenter was that of a snail common to Damba Island in Lake Victoria. The snail was able "to keep the ants at a distance by surrounding itself with bubbles of foam into which the ants could not penetrate" (Carpenter 1920, p. 280). I have seen, in Ivory Coast, a snail react in similar fashion when attacked by workers of the formidable ponerine ant *Paltothyreus tarsatus*. The froth effectively stemmed the attack. My field notes state: "One *P. tarsatus* worker was walking (staggering) away and was glistening with the snail-produced fluid; the ant did not seem to have full use of its antennae, mandibles, etc."

Defensive behavior in some ant species develops specifically in response to army ants. Field and laboratory observations revealed that the ant *Novomessor albisetosus* (subfamily Myrmicinae) responds to attack by *Neivamyrmex nigrescens* with two "clusters" of behavior: "escape with or without the brood, and aggressive defense" (McDonald and Topoff 1986, p. 351). These behaviors, especially escape, are specific to *N. nigrescens* attack and emerge in *N. albisetosus* workers that have no previous experience with army ants. Aggressive defense, at least under laboratory conditions, includes recruitment of defenders, approach, and attack and is the preferred method of defense. Only when army ants gain access to the nest does brood evacuation (escape) commence. However, *N. albisetosus* workers do not respond to army ant attack with aggressive defense until they are seven to nine weeks of age. Thus aggressive behavior is a temporal phenomenon. In fact, Philip McDonald and Howard Topoff (1986) noted that all these defensive behaviors develop with age in workers, even though they have no experience with army ants. Evidence suggests that the course of this development is altered when *N. albisetosus* workers gain such experience.

Some *Camponotus* species (subfamily Formicinae) also respond in specific ways to army ant attack (LaMon and Topoff 1981, p. 1070). In Arizona, three species of this genus—*C. festinatus, C. ocreatus,* and *C. vicinus*—responded to *Neivamyrmex nigrescens* with two distinct patterns of alarm and defense: aggressive nest defense and quick nest evacuation and dispersal (LaMon and Topoff 1981). Aggressive nest defense commonly involved the recruitment of major workers to the nest entrance. Nest evacuation entailed an exodus of the entire colony, including eggs, larvae, pupae, and immature callow workers, which were carried from the nest in the mandibles of the workers. These response patterns did not occur in disturbances not involving army ants.

Rapid and complete nest evacuation was *C. festinatus's* only response to contact with army ants. Brood-laden workers poured from the nest and climbed to the tops of adjacent vegetation, where they remained, motionless, for several hours before gradually returning to the nest. *C. ocreatus* and *C. vicinus,* on the other hand, exhibited a more graded response involving primarily major workers. These workers were recruited tactiley by sister workers to aggressive nest defense, emerging from the nest to patrol within 1 or 2 meters of the entrance, in the case of *C. ocreatus,* or to remain within 20 cm of the entrance in the case of *C. vicinus.* The major workers of these two species constitute a soldier caste, and Brent LaMon and Topoff (1981) hypothesized that the evo-

lution of colony defense systems and polymorphism has been significantly shaped by interspecific ant predation. The aggressive defensive
role of major workers has also been observed in the myrmicine ant *Pheidole oxyops,* which responds strongly to the army ant *Labidus praedator*
(Fowler 1984). LaMon and Topoff (1981) could not explain why *C. festinatus* resorted to immediate nest evacuation to the exclusion of aggressive nest defense.

The myrmicine ant *Pheidole desertorum,* like *C. festinatus,* also reacts to
Neivamyrmex nigrescens attack by evacuating its nest. In this species,
brood evacuation occurs in two distinct phases. When army ants are
approaching but are still relatively far from the nest, workers evacuate
the brood from the interior of the nest but remain massed at the nest
entrance. Only when the army ants draw nearer do the brood-laden
workers run away (Droual 1984). Curiously, *P. desertorum* conducts frequent emigrations between nests belonging to a cluster. Only one nest
is occupied at a time, and emigrations are not normally stimulated by
army ant attack. Robert Droual (1984) concluded that these emigrations
might nevertheless represent a defense strategy against army ants because a fleeing colony can quickly find and enter a previously occupied
nest and in doing so enhance the survival of brood and winged males
and females.

A fascinating way prey ant species may circumvent army ant predation was suggested by Nigel Franks. In an examination of prey ant population composition in the wake of army ant raids, Franks (1982b)
discovered that incipient colonies of prey ants occur in greater abundance in the old swarm paths of *Eciton burchelli* than in areas not recently
raided. Two hypotheses might explain this greater abundance of new
colonies of prey ants in recently raided areas: (1) founding queens are
more successful in raided areas because there are fewer established conspecific colonies present with which to compete; or (2) queens may actively seek recently raided areas (perhaps by detecting army ant trail
pheromones), thereby avoiding competition with conspecific colonies
and also reducing the probability that the colony will encounter army
ants in the near future (*E. burchelli* colonies exercise foraging and movement patterns that lower the probability that they will forage over their
own previous paths). Franks's data "raise the possibility that the queens
of the ant species most heavily preyed upon by *Eciton burchelli* may also
be able to elude their predators by founding their colonies in the aftermath of army ant raids" (Franks 1982b, p. 278).

Typically, when attacked by army ants, social wasps such as *Polistes
erythrocephalus* abandon their nests to the ants, which plunder them of
larvae and pupae with great dispatch (Young 1979). Although social

wasps are often aggressive and are larger than the attacking army ants, their strategy of "escape first" prevails. The rapidity with which some species flee suggests that they can detect the threat of attack in the absence of direct contact with the ants and that the wasps identify the army ants with some degree of specificity (Chadab 1979). This is certainly the case with the small (about 5 mm) social wasp *Protopolybia exigua.*

A member of the family Vespidae, subfamily Polistinae, *P. exigua* constructs a nest 4 to 12 cm long, of a single horizontal comb enclosed by a paper carton envelope and attached to the underside of a leaf (Chadab 1979). There is but a single, small entrance to the nest. Ruth Chadab (1979, p. 117) discovered that when army ants, notably *E. burchelli* and *E. hamatum,* approached the nest, several to 20 or more wasps gathered on the nest surface and its adjacent leaf and began to fan their wings intermittently and to produce a buzzing sound, a behavior pattern called group fanning (Figure 6.3). This alarm response is produced in synchronized pulses. Each buzz is about 0.9 second long followed by a 1.3-second pause. The tip of each wasp's gaster is pressed to the substrate, with the head and front legs lifted, vibrating the nest envelope and presumably communicating alarm to wasps within the nest.

The group fanning continued until the first three or four army ants ran onto the nest. At the encroachment of the ants, the wasps flew in unison from the nest. Chadab (1979) determined experimentally that the wasps detect army ants by sight and odor, an early warning system that permits speedy evacuation from the nest once physical contact with the ants occurs (Figure 6.4). Although the wasp brood could not be rescued, the adult population of the nest escaped to recolonize and start anew.

Angiopolybia pallens, another Neotropical wasp, signals alarm with a very audible din. The wasps rush from the nest entrance, which is situated at the end of the long downward spout, and soon cover most of the envelope. They turn face downward and tap on the nest with their mandibles (Chadab-Crepet and Rettenmeyer 1982), at the same time flipping their wings rapidly and tapping the tips of their gasters against the nest. This behavior is quickly synchronized, with the wasps moving in unison to produce a distinct buzzing or rattle, a chorus of sorts. In large colonies, 200 or more wasps may participate in this behavior, and the sound of the alarm can be heard at a distance of 8 meters (Chadab-Crepet and Rettenmeyer 1982).

Although escape and nest abandonment are the most common reactions of wasps to army ant attack, some species employ, at least initially, an aggressive defense strategy. Large wasps such as *Apoica pallens* and *Polybia velutia* attack approaching army ants. They grasp the ants behind

Figure 6.3. The nest of the small social wasp *Protopolybia exiqua* on the underside of a palm leaflet. The wasps on the nest surface are group fanning in response to the approach of foraging army ants. (Photograph by Ruth Chadab-Crepet. Reprinted, by permission of Ruth Chadab-Crepet, from Chadab 1979.)

the head with their mandibles, fly off, and drop them several meters away. If the raiding ant column is a weak one, this defense might save the wasp colony, but in the face of massive recruitment and attack, ant removal is futile (Chadab-Crepet and Rettenmeyer 1982). Other wasps—*Synoeca septentrionalis*, for example—plug the nest entrance with their heads. With persistence, however, attacking army ants are able to pull the blockading wasps by their antennae from the nest. Once the nest entrance is breached, the demise of the wasp colony is assured.

The swarm-founding social wasp *Angelaia yepocapa* becomes extremely agitated when attacked by army ants and will sting anyone or anything that passes close to the nest, including dedicated researchers (O'Donnell and Jeanne 1990). Most of the several thousand individuals in one colony under attack clustered on a branch near their cavity-bound nest, while a few individuals landed on the ground, oriented to individual ants, and

Figure 6.4. The nest of a wasp of the genus *Polybia* being plundered by raiding workers of *Eciton hamatum*. The adult wasps evacuated the nest, abandoning their brood to the ants. (Photograph by Carl W. Rettenmeyer, Connecticut Museum of Natural History.)

pecked at them with their mandibles, sometimes striking them. These resolute individuals were ultimately overwhelmed and destroyed by the foraging ants (O'Donnell and Jeanne 1990).

Predation by ants is considered a substantial selective force in the evolution of nest architecture in tropical social wasps (Jeanne 1975). Most nests are designed to limit access to the brood. Some paper nests are suspended from narrow petioles to which wasps apply an ant-repelling secretion (Jeanne 1970). Other nests consist of combs housed within a paper envelope (Figure 6.3). Little is known of the role of army ants as a direct selective force on the evolution of nest architecture. Perhaps their influence is minimal, for the group attacks of army ants generally appear undeterred by nest structure (Jeanne 1970).

Tropical social wasps that nest in association with other species of wasps and ants appear most successful in avoiding army ant predation. Such associations represent but a few of the nesting and egg-laying associations found among aculeate Hymenoptera, birds, and reptiles in the tropics (Gotwald 1984).

Chadab-Crepet and Rettenmeyer (1982, p. 273) found that the only consistently effective defense against army ants "is for the wasps to live in trees occupied by certain species of dolichoderine ants in the genus *Azteca*. These tiny ants mob any ant attempting to walk on their tree. The first advancing army ants are each pinned down by 50–100 *Azteca* eliminating the possibility that the ants will recruit additional followers. Wasps which nest in *Azteca* trees thereby receive protection from army ants."

E. A. Heere and associates found numerous wasp species of the genera *Mischocytarus, Polistes,* and *Polybia,* among others, nesting in association with *Pheidole* ants on *Marieta poeppigii* (Melastomataceae), and *Allomerus octoarticuatus* ants on *Tococa guianensis* (Melastomataceae). Further investigation showed that *Eciton burchelli* and *E. rapax* avoid climbing on those species of plants, in sharp contrast with their rapid examination of other plants in their path (Heere et al. 1986).

Termites, as noted previously, also number among the social insects preyed upon by army ants. But termites appear well prepared for defense. At critical times they marshal their soldiers, a caste commonly characterized by heavily sclerotized heads, powerful adductor muscles, and sharp, biting mandibles (Wilson 1971a). The mandibulate soldiers of some species also employ a chemical defense, often salivary fluids that are either toxic or a gluelike substance that entangles their victims. Among certain Rhinotermitidae and Termitidae, the soldiers possess a modified frontal gland from which an entangling, incapacitating fluid can be sprayed (Wilson 1971a). Termites possess a truly impressive defensive arsenal.

In Africa, army ants of the genus *Dorylus,* especially the hypogaeic subgenus *Typhlopone,* rank as the most important predators of termites (Bodot 1961, 1967; Abe and Darlington 1985; Darlington 1985), but few observations of the termites' defensive behavior have been recorded. This is not surprising, given the cryptic nature of termite biology. Darlington (1985) witnessed aspects of one such attack on the fungus grower *Macrotermes michaelseni* by D. (*T.*) *juvenculus* near Kajiado, Kenya. In this region of Kenya, *Dorylus* raids are responsible for 80 percent of the mortality of mature *Macrotermes* nests (although total nest mortality is low). Since the death of many of these nests is often quite sudden, Darlington assumed that the actual fighting was of short duration.

The *Macrotermes* nest for which an attack was recorded consisted of two adjacent conical mounds of earth. Near one of the mounds, a large number of termites, not including soldiers, gathered, fully exposed, dispersed over the soil surface and vegetation. These individuals were attacked by the ants and were carried back inside the mounds. They apparently offered little or no resistance. Much of what occurred within the nest could only be inferred. Actual defense by termite soldiers was not seen.

The nest itself, excavated during the onslaught, contained piles of decomposing termites. The fungus combs had been stripped of their spherical white conidiophores, the vegetative fruiting bodies (later observations revealed that they were indeed harvested by the ants). The termite king and queen were still alive but had been moved to an improvised cell above the royal cell. There was no indication that the termites had attempted to wall themselves off with earth, their common practice when a wall of their nest is damaged. Surviving termites exposed during the excavation were immediately attacked by the ants. Neither soldiers nor worker termites attempted to defend themselves (Darlington 1985).

Excavation of this and several other moribund nests revealed a defensive strategy that included plugging the small entrance holes in the roof and walls of the royal cells with soil and, in some instances, removing the reproductives from the royal cell to other parts of the nest. *Dorylus* is obviously successful in its assaults on termite nests, despite the defensive endowments of the termites. Do army ants ever lose the battle? That is difficult to say. Although army ant head capsules have been found in rubbish deposits in *Macrotermes* nests, Darlington (1985) could determine no way of discovering the frequency of failed doryline attacks.

The antipredator strategies of even some small vertebrates may have been shaped by army ant predation. The long legs and leaping ability of certain African shrews, for instance, appear not to be related to the pursuit of prey as much as to escape from driver ant attack (Brosset 1988).

Role of Army Ants in Maintaining
Species Diversity

The importance of natural disturbance in the maintenance of high species diversity among plants in tropical moist forest is a tenet of tropical ecology. Limited destruction of the forest canopy from tree fall, for instance, produces gaps that are colonized by propagules and pioneer species different from those of the surrounding forest. Consequently, a forest is a mosaic of gap-created patches of plants in different stages of maturity and succession (Whitmore 1978). The size of the gap, which affects its microclimate, influences species composition and spatial arrangement in rain forest, since different species are successful in gaps of different sizes (Whitmore 1978, Denslow 1980). The frequency with which gaps are created also figures significantly in species diversity. The continuous production of gaps suggests that tropical communities are never in equilibrium and that maximum diversity is maintained by intermediate levels of disturbance (Connell 1978).

Predators, including seed predators, also influence diversity in the tropics (Janzen 1969, 1970; Baker 1970). The so-called predation hypothesis maintains that predators (and parasites) influence biological diversity by preventing any one prey species from building its population size to the point where it monopolizes the resource supply (Baker 1970). Army ants, as predators of social insects, increase or at least maintain species diversity among ground-dwelling ants (Franks and Bossert 1983). But army ants can also be seen as analogues of physical disturbance that produce gaps in the forest floor ant fauna and thereby create opportunities for new colonies of both prey species and nonprey species to establish themselves.

Field studies and simulation models led Franks and Bossert (1983) to conclude that areas recently raided by *E. burchelli* have a higher diversity of ant species, and that as these areas age the ant community is subject to succession. In a field experiment involving the placement of artificial nests in recently raided areas and in control areas, incipient colonies, especially of *E. burchelli* prey, occurred in greater abundance in the raided areas. Franks and Bossert (1983, p. 159) proposed that "in the aftermath of swarms when prey colonies have been killed or cropped to smaller size, intraspecific predation of queens of prey species may be reduced, with the result that incipient nests of these species can be founded in greater numbers."

The succession of ant species that follows in the wake of *E. burchelli* swarm raids often features the rapid increase of an opportunistic nonprey ant species of the genus *Paratrechina* (subfamily Formicinae). This species may be responsible for the greater turnover of *Pheidole* species in raided areas than in nonraided areas. Together, *Paratrechina* and *Pheidole* are more successful than prey species at quickly capitalizing on the transitory opportunities provided by army ant raids (Franks and Bossert 1983). These small ants outcompete the large-bodied prey species, which belong mostly to the subfamilies Ponerinae and Formicinae.

Economic Impact

Army Ants as Beneficial Insects

Beneficial economic effects of ants, relative to human ventures, may be derived at several levels in a hierarchy of relationships that ranges from general benefits to increasingly specific ones. The general, or pandemic, benefits originate with ants' roles in helping to create environments hospitable to human endeavors (Gotwald 1986). Ants are of general value, for example, because they move and aerate the soil and facilitate the cycling of nutrients. They are of direct benefit when they are consumed as human food.

Ants may also be of direct value in agroecosystems. For instance, predaceous ant populations can be manipulated to control agricultural pests, a practice that antedates modern intregrated pest management and biological control by more than a millennium. Most likely, the first practical use of predatory ants to control agricultural pests was recorded by the Chinese in A.D. 304 (Way and Khoo 1992). Later, in the ninth- and tenth-century book *Ling Pio Lu Yi* (Wonders from south China), Liu Shun described how fruit growers placed nests of the Asian weaver ant, *Oecophylla smaragdina,* in their citrus trees, deliberately encouraging the ants to reside in their orchards. Supposedly, this predatory ant, known in Chinese as *huang ji yi* (yellow fear ant), attacks and dines on citrus pests (Liu 1939, DeBach 1974, Huang and Yang 1987). It is a practice still in use (Huang and Yang 1987). Today, the possibility of manipulating ant populations to effect some measure of pest control, especially of canopy species in such crop trees as cocoa, may be an important consideration in designing integrated pest management programs (see Majer 1974, 1976a,b,c, Way and Khoo 1992).

Army ants, notorious predators, are obvious candidates in the search for beneficial species. The literature is sprinkled with anecdotal accounts of their beneficent accomplishments. In Angola, F. C. Wellman (1908, p. 226) considered *Dorylus* (*Anomma*) *nigricans* economically important: "Careful housewives welcome the approach of the ants and joyfully vacate for them the bungalow," for "after a column of 'army ants' has minutely explored a dwelling not a bug, beetle, cockroach, mouse, rat, snake, or other pest remains behind."

L. Burgeon (1924a) regarded the driver ant *D. (A.) wilverthi* to be of general value because it destroys agricultural pests. Other researchers (Alibert 1951, Strickland 1951) speculated, without supporting evidence, that driver ants are valuable on cocoa farms because they kill insect pests. A similar, and equally unsubstantiated, claim was made for *Labidus coecus* and *L. praedator* on Brazilian cocoa farms (Delabie 1990). A preliminary investigation (Gotwald 1974b) of the behavior of driver ants on cocoa farms in Ghana revealed that the ants do not forage in the cocoa tree canopy where most insect pests are found (including ants that harbor harmful homopterous insects). Dominant canopy ant species, such as *Oecophylla longinoda,* may successfully repel driver ants. The absence of driver ants as predators of the cocoa canopy fauna suggests that they are not of direct economic importance in determining the composition of that fauna; therefore, driver ants appear to have little potential value in the biological control of cocoa pests (Gotwald 1974b).

In the American Southwest, the role of ants as predators of the screw-worm, a livestock pest, was investigated by A. W. Lindquist (1942). The screw-worm (*Cochliomyia hominivorax*), a fly of the family Calliphoridae, lays its eggs in wounds or in the nostrils of its hosts—sheep, goats, and cattle. Newly hatched larvae burrow into the host. When mature, the carcass-infesting larvae leave the host animal, either when the host is alive, in which case the larvae drop to the substrate and are dispersed over a wide area, or after the host is dead, and the larvae concentrate beneath the host carcass.

Lindquist (1942), observing the emergence of the screw-worm larvae from their hosts, noticed that large numbers of ants quickly appeared and attacked and killed the larvae. Among these predator ants, the most frequently observed was the ecitonine *Labidus coecus*. In controlled experiments, a 4.1 percent emergence rate of adult flies occurred when infested host carcasses were exposed to ants, while a 93.1 percent emergence rate was recorded from carcasses protected against ant attack. Ants therefore constitute an important natural control of the screw-worm, and *L. coecus* is the most important species among them. Although the screw-worm has since been controlled by other methods, Lindquist's research illustrates the positive role army ants may play in animal husbandry and agronomics.

Army ants have been credited with killing pest insect species in a variety of agroecosystems. *Dorylus (D.) helvolus,* for example, attacks the larvae of the maize stalk borer, a moth at one time considered the most important pest of maize in South Africa (Moore 1913). In Kenya, I have observed the driver ant *D. (A.) molestus* killing caterpillars on maize. *D. (D.) helvolus* has also been credited with having a beneficial effect on crops and pastures in southern Africa by reducing herbivore populations (Prins et al. 1990). Likewise, Burgeon (1924a, p. 65) was certain that *D. (A.) wilverthi* destroyed a multitude of harmful insects. And G. R. Dutt (1912) remarked that *D. (Alaopone) orientalis* was beneficial in India because it attacked and killed in large numbers the ant *Pheidole indica*, which he described as an occasional nuisance. In Brazil it was expected that *Labidus praedator* would be an effective control agent of the corn weevil in stored corn. The ant, however, did not live up to its reputation (Caetano 1991).

Army Ants as Pests

Army ants are rather benign creatures, from the human point of view, notwithstanding the few fictionalized portrayals that ascribe to these ants diabolic intent (see Chapter 7). To be bitten by these ants is not pleasant, but neither is it life threatening. And while some people might welcome to their domicile a vermin-expelling visit by a famished, marauding army ant colony, most would consider such a visit, and their subsequent eviction, an unprovidential annoyance, even an affront. But only one species, *Dorylus (Alaopone) orientalis*, found in India, Nepal, Sri Lanka, southern China, and Burma (Wilson 1964), ranks as a serious agricultural pest. It is the only army ant that regularly includes plants in its diet, and it attacks a variety of domesticated plants (see Table 6.1).

The role of *D. (A.) orientalis* as an agricultural pest is now irrefutable, although it was not always so. E. Barlow (1899), the first to report this ant as a plant pest, wrote that it fed on potatoes and produced significant damage. The eminent French myrmecologist Auguste Forel's argument that all army ants are decidedly carnivorous (quoted in Barlow 1899) drew a defiant response from entomologist E. E. Green (1903, p. 39):

With all due deference to Dr. Forel's acknowledged learning on the subject of ants, I most emphatically contradict this statement, as far as it refers to *Dorylus orientalis*, West. The workers of this species live entirely underground, and I can assert from repeated personal observation, that they are most confirmed vegetarians. I found it quite impossible to grow potatoes in Pundalnoya [India], solely on account of this insect, and they were most aggravating in their systematic attacks upon the tubers of dahlias, and the roots of the common sunflower (*Helianthus* sp.). In the case of tubers they form galleries through and through the substance, and in the case of roots they eat off the tender bark below the collar. I

Table 6.1. Cash crop plants, garden vegetables, and ornamentals attacked and damaged by *Dorylus* (*Alaopone*) *orientalis*

Plant	Country	Reference
Artichoke	India, Sri Lanka	Lefroy 1906, 1909 (in Roonwal 1975)
Beans	China	Chen and Xie 1991*
Cabbage	India, Sri Lanka, China	Lefroy 1906, 1909 (in Roonwal 1975); Chen and Xie 1991*
Carrots	Sri Lanka	Hutson 1933a (in Roonwal 1975)
Cauliflower	India, Sri Lanka	Lefroy 1906, 1909 (in Roonwal 1975)
Citrus	Sri Lanka	Hutson 1933b (in Roonwal 1975)
Coconut palm	India, Sri Lanka	Fletcher 1919; Hutson 1939 (in Roonwal 1975)
Cornflower	India	Stebbing 1905
Dahlia	Sri Lanka	Green 1903
Eggplant	China	Chen and Xie 1991*
Ginger	Sri Lanka	Hutson 1937 (in Roonwal 1975)
Groundnut (peanut)	India	Fletcher 1920 (in Roonwal 1975); Rajagopal et al. 1990; Mahto 1991
Kohlrabi	Sri Lanka	Rutherford 1914 (in Roonwal 1975)
Mango	India	Menon and Srivastava 1976
Muskmelon	India	Musthak 1981 (in Rajagopal et al. 1990)
Onions	Sri Lanka	Hutson 1933a (in Roonwal 1975)
Peppers	China	Chen and Xie 1991*
Potatoes	India, Sri Lanka	Barlow 1899; Green 1903; Stebbing 1905; Roonwal 1975
Sugarcane	Sri Lanka	Fletcher 1914 (in Roonwal 1975)
Sunflower	Sri Lanka	Green 1903
Sweet potatoes	China	Chen and Xie 1991*
Tree tomato	Sri Lanka	Hutson 1939 (in Roonwal 1975)
Turnips	China	Chen and Xie 1991*
Watermelon	China	Chen and Xie 1991*

*Unpublished data supplied by Chen Yi, Zheijiang Agricultural University, Hangzhou; and Xie Fuyi, Agricultural School, Xiangxi, Hunan, People's Republic of China.

have made very careful observations on the point and have completely satisfied myself that the *Dorylus* was really feeding upon the vegetable tissues, and was not merely hunting for another insect.

Except for the damage it does to partially mature mango fruits (Menon and Srivastava 1976) and eggplant, tomato, and watermelon (Chen Yi and Xie Fuyi pers. comm. 1991), *D.* (*A.*) *orientalis* mostly confines its attacks to the roots and tubers of plants (Roonwal 1975). Of most recent concern is the damage this species does to groundnut (peanut) crops in southern India. The workers bore through the tender subterranean pods of this plant and feed on the developing kernels (Figure 6.5). When harvested, many of the attacked pods are empty but for a black powder,

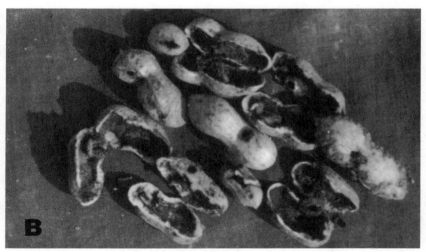

Figure 6.5. Groundnut pods damaged by *Dorylus* (*Alaopone*) *orientalis*, as they appear freshly uprooted in the field (A) and cut open (B). (Photographs by Y. Mahto. Reprinted, by permission of Y. Mahto, from Mahto 1991.)

evidence of a secondary fungal infection (Rajagopal et al. 1990, Veeresh 1990). Researchers attempting to combat the ant with pesticides have found that carbofuran controlled-release granules are most effective in reducing pod damage (Rajagopal et al. 1990). The numerous groundnut varieties vary in their susceptibility to *orientalis* attack. Some varieties appear immune, while others sustain up to 45 percent damage (Mahto 1991).

D. orientalis is also an enemy of the honey bee in India, for it attacks hives and devours both adult and larval bees (Ghosh 1936). Clearly, no other army ant species enjoys the notoriety of *D. orientalis* for the damage it does to cash crops, garden vegetables, and ornamentals.

One other species noted for its negative economic impact is *D. (D.) helvolus* in southern Africa. This species preys intensively on the pupae of *Cactoblastis cactorum*, a moth of the family Pyralidae that is used in the biological control of cactus weeds (Prins et al. 1990). By destroying up to 45 percent of the moth's pupae, *helvolus* prevents its effective use in control programs. Farmers in southern Africa have reported that *helvolus* also damages potato and dahlia tubers, not unlike its congeneric "cousin," *D. (A.) orientalis*.

Army ants play a tangled, multidimensional role in tropical ecosystems; they affect a large number of organisms in a variety of ways. To remove them from an ecosystem is to remove the violins from an orchestra: it is to simplify the harmonies. This view is not overstated speculation. When tropical rain forest is fragmented, army ants and their avian followers are among the first organisms to disappear (Lovejoy et al. 1984, 1986). The loss of species from isolated forest fragments is now understood as a process of ecosystem decay. Documenting and comprehending ecosystem decay is prerequisite to determining the minimum critical size for isolated remnants of tropical forest that will maintain communities close to their characteristic diversity (Lovejoy et al. 1983).

Because army ant foraging is cyclical, a colony is a useful source of insects for ant-following birds only during its nomadic phase. Therefore, each forest fragment must be able to support enough colonies, perhaps a minimum of three, so that at least one will be in the nomadic phase at any given time (Lovejoy et al. 1986). If *Eciton burchelli* requires about 30 hectares of continuous forest, as it apparently does on Barro Colorado Island (Franks 1982a), then a 100-hectare forest fragment or reserve could theoretically sustain a representative ant-following bird community (Lovejoy et al. 1986).

In Mexico, destruction of rain forest has produced extensive fragmentation that has significantly depleted colonies of *Eciton* and *Labidus*

(Coates-Estrada and Estrada 1989). "The disappearance of these ants and of their contribution to the sustaining of important ecological links through the potential food they make available to many understorey vertebrates," wrote Rosamond Coates-Estrada and Alejandro Estrada (1989, p. 291), "is an important conservation problem and will seriously hamper our understanding of the ecological role played by army-ants as 'keystone' organisms."

7

Myth and Metaphor

The army ant is a noteworthy example of an organism whose popular reputation was shaped at the interface of science and mythology, at the penumbral border between fact and fiction. Although richly deserving of scientific wonderment and attention, the army ant and its reputation were initially fashioned more in the imaginations of mythmakers than through the analysis of data carefully gathered by scientists. But is the hypothesis—the scientist's presupposition that explains an observable phenomenon—any less creative in its origin than the fictional constructs of writers and artists? Imaginative hypotheses, after all, have generated much of the published quantitative research on army ants, and a host of other creatures as well. Undeniably, our knowledge of army ant behavior and ecology has emerged from the creative energy invested in the articulation of hypotheses and the inventive design of field and laboratory studies. In early explanations of army ant behavior, the distinction between hypothesis and fantasy was tenuous indeed.

The first descriptions of army ants and their plunderous proclivities came from naturalists, trained and untrained. Even the most sophisticated of these observers was occasionally given to exaggeration, providing fuel for the fires of mythology. The indigenous peoples of the New and Old World tropics had their own tales to tell about the exploits of army ants, but these, some with the hint of allegory, remained the cultural property of those people. Their influence on the observations of European and American naturalists was minimal.

The great temptation to which many naturalists succumbed when recording their observations of army ants was to see in them the regimented movements and destructive force of human armies. The anthropomorphizing of army ant behavior was both helpful and counterproductive in achieving an accurate understanding of the proximate and ultimate forces that have guided army ants through evolutionary time and contemporary space. This legacy—that is, the portrayal of army ant behavior using military metaphor—is a permanent part of the scientific literature.

The layperson's indelible impression of army ants has been conditioned more by illusion than reality. Not that the layperson encounters frequent images of army ants, but the few depictions that do confront the uninitiated constitute a tour de force that exploits a common, largely subliminal, entomophobia. Hordes of insects laying waste to plants, animals, and humans alike is a powerful image that penetrates the primeval core where ancient human fears reside.

Army Ants in the Eyes of Naturalists

The predatory antics of army ants clearly did not escape the attention of eighteenth- and nineteenth-century explorers, adventurers, and naturalists. The written observations of such journeyers are spiced with accounts of army ants on the march. William Swainson (1835, p. 100), for instance, in his discourse on the geography and classification of animals, cited the comments of a Mr. Smeathman, "who lived many years" in Sierra Leone. One species of ant, "which seems at times to have no fixed habitation," Smeathman related, "ranges about in vast armies. . . . By being furnished with very strong jaws, they can attack whatever animal impedes their progress; and there is no escape but by immediate flight, or instant retreat to the water." There can be little doubt that the ant referred to here belongs to the Old World army ant genus *Dorylus*.

But the award for hyperbole in describing army ants surely belongs to that intrepid American explorer Paul B. Du Chaillu. In a detailed account of his exploration of Equatorial Africa, Du Chaillu (1861, p. 359) referred to the "bashikouay" ant, which, he stated "is the most voracious creature I ever met. It is the dread of all living animals from the leopard to the smallest insect." He went on to say (pp. 359–360):

It is their habit to march through the forests in a long regular line—a line about two inches broad and often several miles in length. All along this line are larger ants, who act as officers, stand outside the ranks, and keep this singular army in order, [and] when they get hungry the long file spreads itself through the forest in a front line, and attacks and devours all it comes to with a fury which is quite irresistible. The elephant and gorilla fly before this attack. The black men run for their lives. Every animal that lives in their line of march is chased. They seem to understand and act upon the tactics of Napoleon, and concentrate, with great speed, their heaviest forces upon the point of attack. In an incredibly short space of time the mouse, or dog, or leopard, or deer is overwhelmed, killed, eaten, and the bare skeleton only remains.

In the New World tropics, Henry Walter Bates, a prodigious collector of biological specimens, recorded detailed observations of army ants, especially of the genus *Eciton*. Bates, an Englishman who landed in Amazonas in 1848 in the company of the not-yet-famous Alfred Russel Wallace, remained in South America for eleven years. His long stay no doubt accounts for the fact that he returned to England with 14,712 species, 8000 of which were new to science. Although Bates was painstakingly scientific in his descriptions of the flora and fauna of the Amazon Basin and was not given to bravado, he too noted the rapacity of army ants. In his well-known book, *The Naturalist on the River Amazons* (1863, pp. 371–372), Bates described an encounter with army ants:

When a pedestrian falls in with a train of these ants, the first signal given him is a twittering and restless movement of small flocks of plain-coloured birds (ant-thrushes) in the jungle. If this be disregarded until he advances a few steps farther, he is sure to fall into trouble, and find himself suddenly attacked by numbers of the ferocious little creatures. They swarm up his legs with incredible rapidity, each one driving his pincer-like jaws into his skin, and with the purchase thus obtained, doubling in its tail, and stinging with all its might. There is no course left but to run for it; if he is accompanied by natives they will be sure to give the alarm, crying "Tauoca!" and scampering at full speed to the other end of the column of ants. The tenacious insects who have secured themselves to his legs then have to be plucked off one by one, a task which is generally not accomplished without pulling them in twain, and leaving heads and jaws sticking in the wounds.

Another naturalist of the New World tropics, Thomas Belt (1874, p. 18), roamed about Nigaragua. He recalled how one small species of army ant there would "visit our house, swarm over the floors and walls, searching every cranny, and driving out the cockroaches and spiders, many of which were caught, pulled or bitten to pieces, and carried off." About another species Belt (1874, p. 21) wrote: "Crickets, grasshoppers, scorpions, centipedes, wood-lice, cockroaches, and spiders are driven out from below the fallen leaves and branches. Many of them are caught by the ants; others that get away are picked up by the numerous birds that accompany the ants, as vultures follow the armies of the East. The ants send off exploring parties up the trees, which hunt for nests of wasps, bees, and probably birds. If they find any, they soon communicate the intelligence to the army below, and a column is sent up immediately to take possession of the prize."

The twentieth century certainly possesses its own school of army ant observers, a diverse collection of biologists, writers, and travelers. Although most of their observations seem rather grim in comparison with the florid style of the 1800s, some found humor in the confrontations between humans and army ants. A. W. Cardinall (1927, p. 185), onetime district commissioner of the Gold Coast, for instance, reminisced: "A short time ago the D.C. at Ejura witnessed an amusing incident. Three white men were his guests and he had left them on the verandah for a moment or two. On his return he found them stark naked, swearing and busily picking off ants. From personal experience I know that nothing short of rapid undressing can deal with an attack. The ants will soon leave one's empty clothes."

Few indeed are the tropical biologists who have not inadvertently stepped into an agitated mass of army ants and had to drop their trousers with the hurried dispatch of the falling curtain at the conclusion of a bad play.

Arthur Loveridge (1949), a herpetologist who lived for many years in East Africa before becoming a curator at the Museum of Comparative Zoology at Harvard University, could not ignore his encounters with siafu, the army ants commonly known in English as driver ants. He described a siafu invasion of his home near Nairobi: "I was reading that evening when, about nine o'clock, I gradually became aware of sundry small noises which together constituted quite a volume of sound. Taking up the table lamp I went to my bedroom, where an astonishing sight awaited me. The whitewashed walls were a moving mass of siafu. They swarmed upon the books in the bookcase and overran the furniture. It was the movement of their myriad feet which produced the curious sound I had heard" (Loveridge 1949, p. 205).

And G. D. Hale Carpenter (1920, p. 280), an English biologist who, while undertaking a study of the bionomics of the tsetse fly in Uganda, found himself swept up by World War I and conscripted for duty with troops on the Uganda-German (Tanganyika) frontier, wrote: "During the campaign in German East Africa I was twice turned out of the little tent in which I was sleeping on the ground by an invasion of Ensanafu, and had to bolt, and then, after picking off those ants which had already attached themselves to various parts of my person, make a dash for boots, and by frequent painful visits gradually withdraw the bedding and shake down elsewhere outside, while the ants proceeded along the line to some one else's tent, to my secret gratification, for no one likes to be the only one who is turned out in the night!"

Casual incidental remarks about army ants are sprinkled, like seasoning on an already flavorful dish, throughout a rather large body of literature on Africa. Army ants have intruded on the lives of many. In the powerful and memorable classic *Out of Africa*, Baroness Karen Blixen-Finecke, under her famous nom-de-plume Isak Dinesen (1937, p. 35), complained that her dogs had been attacked by these "murderous big ants, the *Siafu*," and that the ants had to be picked from the dogs "one by one."

Theodore Roosevelt (1910, p. 435), twenty-sixth president of the United States and an avid outdoorsman, described his encounters with army ants during a hunting expedition to Africa: "We also saw a long column of the dreaded driver ants. These are carnivorous; I have seen both red and black species; they kill every living thing in their path, and I have known them at night [to] drive all the men in a camp out into the jungle to fight the mosquitoes unprotected until daylight. On another occasion, where a steamboat was moored close to a bank, an ant column entered the boat after nightfall, and kept complete possession of it for forty-eight hours. Fires, and boiling water, offer the only effectual means of resistance."

More recently, Katharine Hepburn (1987, p. 88) reminisced about the on-location filming of the 1951 movie *The African Queen* and how she inadvertently stood in a "procession of ants": "I tore off my clothes. I was literally covered up to my neck with ants—bitten everywhere except hands and neck and face. Those army ants crossed our campsite in as straight a line as possible, and—luckily for us—they went out the other side. Had they decided to stay—it's *we* who would have had to move. As I was wearing high necks and long sleeves, my bites didn't show."

Members of the contemporary press are not exempt from casting army ants in their legendary role of jungle rogues. In a special report to *Time*, Eugene Linden (1992) recounted his experiences in the remote Ndoki region of northern Congo, which is endowed with perhaps the last pristine rain forest in Africa. The 15-day expedition in which he participated was filled with adventure, including, of course, the predictable encounter with army ants.

> After a meal of soup, salami and cookies, I settle in to sleep, wondering whether the dire reports I had heard from the Japanese researchers had overstated the dangers of the area. A few minutes later, I awake feeling an insect on my finger. Flicking it off, I feel another take its place, and

then suddenly thousands of bugs seem to bite me at once. Seconds later, I hear a strangled cry from Karen as she is attacked as well. Stumbling blindly over roots and a massive column of ants, we tear down a path and dive into the river. Crushing the ants seems to release some chemical distress signal: as we emerge from the river, the aggressive creatures drop on us from everywhere.

Army Ants in Folklore

Whatever their local name—*tauoca* in Amazonas, *tepeguas* in Mexico, or *bashikouay, ensanafu, siafu,* and *kelelalu* in Africa—army ants and tales of their foraging activities enrich the oral traditions of the indigenous peoples of tropical forest and savanna. In some instances, army ants have become, as well, utilitarian items in the daily lives of these same peoples.

The Ashanti of Ghana, the West African country formerly known as the Gold Coast, articulate their respect for *nkran,* the conspicuous driver ant of both grassland and forest, in an anecdote about the hungry python (Gotwald 1982). They say that before the python takes an immobilizing meal, perhaps a duiker (a small antelope of the forest) or similar creature, it will diligently circle the area in search of *nkran.* Should the python discover *nkran,* it will prudently forgo its meal and wait for a more propitious—that is, less dangerous—time to have lunch. Clearly, the Ashanti had observed pythons engorged with food, and consequently lethargic and immobile, inundated by swarming driver ants and no doubt picked clean to the bone. (I have personally seen the skin and bony remains of caged snakes that met a similar fate—screened cages do not deter hungry army ants.)

In Ghana and Togo, the Ewe are said to welcome the driver ants in their fields of maize and cassava as assurance that there are no deadly snakes about. Likewise, a chance visit of army ants to one's hut or house may not be the calamity such a visit might suggest. G. A. Perkins (1869, p. 361), a nineteenth-century visitor to West Africa, reported that "there are times when their visits are most welcome. On their approach every kind of vermin is seized with consternation, and seek safety in flight. Centipedes, Cockroaches, scorpions, etc., etc., leave their hiding-places, and are seen seeking places of greater security, only to fall at last into the clutches of their relentless foe, from whom there is no escape." Army ants: nature's primordial exterminators.

Most curiously, perhaps, army ants have been used medically as sutures. Bizarre though it may seem, surgeons in medieval Spain, France, and Italy commonly employed ants in suturing wounds of the small intestines. The practice of suturing cuts with ants occurred in Turkey, India, and parts of South America and Africa (Gudger 1925), although the persistence of the custom in recent times in some of these places is conjectural. In the tropics this procedure utilizes the soldier subcaste of *Eciton* and *Dorylus* army ants, which possess sharp, tonglike mandibles. The Kikuyu of Kenya make use of army ants in this way (Murray-Brown 1973). The late Alex Haley (1976, p. 12), in his renowned book *Roots,* provided a fictional account of the process: "But when she saw Kunta, she sprang up to wipe his bleeding forehead. Embracing him tightly, she ordered the other children to run and bring her some kelelalu ants. When they returned Grandma Yaisa tightly pressed together the skin's split edges, then pressed one struggling driver ant after another against the wound. As each ant angrily clamped its strong pincers into the flesh on each side of the cut, she deftly snapped off its body, leaving the head in place, until the wound was stitched together."

One last note of interest: Accra, the name of the capital city of Ghana, may have been derived from the Akan word *nkran* through a former spelling, *akra* (Vigah 1977). On casual observation, at least, the influence of army ants on the lives of tropical peoples appears pervasive.

Army Ants and the Military Metaphor

The behavior of army ants surely invites comparison to military maneuvers, strategy, and logistics. Even the earliest references to these ants (e.g., Swainson 1835) described how they move about in "armies." Frank M. Chapman (1929, p. 185), a distinguished ornithologist and former curator of birds at the American Museum of Natural History, recounted the movement of a "detachment" of army ants in Panama and noted of their attack on invertebrates of the forest floor: "One expects to hear the blare of trumpets and cries of agony." Perhaps the military metaphor as applied to army ants reached its pinnacle in the popular literature with the publication of William Beebe's article "The Home Town of the Army Ants" (1919). Beebe, an authority on tropical biology and popularizer of scientific research, used numerous military terms to describe the ants and their biology, including *legionaries, warriors, scouts, soldiers, guards, sappers, booty, battalions, scimitars* (referring to their mandibles), *Spartan, squad, expedition, battle,* and even *court-martial.* "Army ants," he wrote (1919, p. 454), "have no insignia to lay aside, and their swords are too firmly hafted in their own beings to be hung up as post-bellum mural decorations, or—as is done only in poster-land—metamorphosed into pruning-hooks and ploughshares."

The military metaphor is fixed in place with the stubborn cement of time, and at least some of the terms are now incorporated into the scientific literature. The late T. C. Schneirla, prominent behaviorist at the American Museum of Natural History, did more than anyone else to ensure that such terms as *raid, booty,* and *bivouac* achieved a certain exclusivity in their use in describing army ant lifeways. Schneirla, who pioneered modern research into the behavior of army ants, especially in the New World tropics, promoted the military metaphor as a useful exercise in distinguishing army ants from the many other species of predatory ants.

Army Ants Mythologized

No one account did more to immortalize army ants and their murderous exploits than Carl Stephenson's remarkable short story "Leiningen versus the Ants." In it, a Brazilian plantation is besieged by an advancing tide of ravenous army ants that indiscriminately devours every living thing in its path. Leiningen, the intractable owner of the plantation, is warned of the approaching horde: "The Brazilian official threw up lean and lanky arms and clawed the air with wildly distended fingers. 'Leiningen!' he shouted. 'You're insane! They're not creatures you can fight—they're an elemental—an act of God! Ten miles long, two miles wide—ants, nothing but ants! And every single one of them a fiend from hell; before you can spit three times they'll eat a full-grown buffalo to the bones. I tell you if you don't clear out at once there'll be nothing left of you but a skeleton picked as clean as your own plantation'" (Stephenson 1940, p. 682).

As the frenzied mass of ants reaches the plantation, Leiningen mounts his horse and rides toward the oncoming swarm, ready to defend life, limb, and property (p. 685): "It was a sight one could never forget. Over the range of hills, as far as eye could see, crept a darkening hem, ever longer and broader, until the shadow spread across the slope from east to west, then downwards, downwards, uncannily swift, and all the green herbage of that wide vista was being mown as by a giant sickle, leaving only the vast moving shadow, extending, deepening, and moving rapidly nearer."

Of course, this is the stuff of which movies are made. Indeed, Hollywood could not resist the temptation to commit to film this epic, primeval battle of man against nature. In 1954, Stephenson's tale was transformed into *The Naked Jungle*, a cinematic adventure starring Charlton Heston and Eleanor Parker. The ants rampaged, stripping the flesh from men and horses unfortunate enough to stumble into the seething

Figure 7.1. A plantation worker in the film *The Naked Jungle* grimaces in horror as army ants swarm over his body.

legion, leaving in their wake impeccably cleaned skeletons, grisly reminders of their insatiable appetite (Figure 7.1). Live ants abound in the film, scurrying about with convincing predatory authority, but under cursory examination they reveal themselves to be carpenter ants of the genus *Camponotus*.

On a more somber note, one cannot help but speculate that the red ants that suffuse the novel *One Hundred Years of Solitude*, the spectral masterpiece by Nobel Prize winner Gabriel Garcia Márquez, one of Latin America's most celebrated writers, are army ants. We are told, for instance, that the sleeping couple, Aureliano and Amaranta Ursula, "are awaked by a torrent of carnivorous ants who were ready to eat them alive" (p. 411). Later, the devastation of human tragedy descends on the reader as Amaranta dies producing a son, the last of the family line. Aureliano, absorbed in his grief and his nostalgia for lost friendships, forgets his newborn son in his reflective preoccupation. His dreaminess is soon shattered: "And then he saw the child. It was a dry and bloated bag of skin that all the ants in the world were dragging toward their holes along the stone path in the garden" (p. 420).

As excessive and embellished as some fictional accounts are, we may find at their core a kernel of truth. Du Chaillu (1861, p. 361), recounting his explorations of central Africa, reported: "The negroes relate that criminals were in former times exposed in the path of the bashikouay ants, as the most cruel manner of putting to death." And when collecting stories about driver ants in Ghana, I was told of an infant who died in the onslaught of a foraging swarm of these ants. Although I could not confirm this story in any official way, it is not a tale that challenges credulity. Ghanaian women traditionally till their family gardens and when doing so are in the habit of depositing their babies on the ground in the shade of nearby trees.

Is it any wonder, then, given the mythology and the reality of army ants, that these insects are popularly regarded today as nature's consummate predators, warriors whose deeds would make the likes of Genghis Khan blush with envy?

Epilogue: In Defense of Army Ants
(and Other Tropical Beasts)

Alfred Russel Wallace (1878, p. 121), English naturalist, contemporary of Charles Darwin, and coauthor of the theory of natural selection, insightfully wrote that "animal life is, on the whole, far more abundant and varied within the tropics than in any other part of the globe, and a great number of peculiar forms are found there which never extend into temperate regions. Endless eccentricities of form and extreme richness of colour are its most prominent features, and these are manifested in the highest degree in those equatorial lands where the vegetation acquires its greatest beauty and fullest development."

What Wallace witnessed and recognized was the richness and diversity of the living forms that characterize the tropics, especially the tropical moist forests of the world. Although approximately 1.7 million species have been described (750,000 of which are insects), some biologists have estimated that the total number actually residing on our planet may exceed 10 million, and one entomologist has proposed that there may be as many as 30 million insect species alone (Wilson 1985d)! If true, this would represent a truly astounding level of diversity. But we have been surprisingly slow to acknowledge the value of species diversity to the survival of tropical ecosystems, and perhaps even to global ecological stability. E. O. Wilson (1985d, p. 700) suggested that "organic diversity has remained obscure among scientific problems for reasons having to do with both geography and the natural human affection for big organisms. The great majority of kinds of organisms everywhere in the world are not only tropical, but also inconspicuous invertebrates such as insects, crustaceans, mites, and nematodes." That diversity is now threatened as never before.

It is estimated that 245,000 km² of moist tropical forest are destroyed or converted to other forms each year. Farming and the harvesting of wood for pulp, timber, and firewood all contribute to the demise of these forests. Theoretically, the entire biome of 9.35 million km² could be eliminated within 38 years. To put it even more graphically, "it is not unreasonable to suppose that the earth is losing around 670 km² of tropical moist forest a day, or an area the size of . . . Massachusetts each month" (Myers 1983). That amounts to 46 hectares per minute!

In the process, many species, including army ants, are in danger of being lost forever; some have already gone the way of trilobites and dinosaurs, before we even knew they were there. But trilobites and dinosaurs were not victims of anthropocentrism. They did not succumb to the damaging demands of *Homo sapiens,* earth's most cultural species. Environmental ethicist Holmes Rolston III (1985, p. 724) maintained that "artificial extinction, caused by human encroachments, is radically different from natural extinction. Relevant differences make the two as morally distinct as death by natural causes is from murder." Indeed, our species has clearly changed the equation of survival. Should we commit ourselves, then, to maintaining earth's biological diversity? Are there abiding reasons why we should be concerned?

I suppose we could list the practical reasons for preserving diversity: avoiding the possible disruption of global weather patterns, preserving potential sources of new medicines, and maintaining a critical source of income from tourism in developing nations are certainly among them. But as Rolston (1985, p. 726) argued, "There is something Newtonian, not yet Einsteinian, besides something morally naive, about living in a reference frame where one species takes itself as absolute and values everything else relative to its utility." I would suggest that there are other eminent reasons, more sublime than pragmatic, to champion biological diversity. Given our self-defined status as earth's most advanced species, moral and aesthetic reasons should be the most compelling of all. If we truly represent the height of sophistication in which we so gladly cloak ourselves, then we should value other living things and their survival not because it is practical or expedient, but simply because it is the right thing to do.

Literature Cited

Abe, T., and J. P. E. C. Darlington. 1985. Distribution and abundance of a mound-building termite, *Macrotermes michaelseni*, with special reference to its subterranean colonies and ant predators. Physiol. Ecol. Jpn. 22:59–74.

Agosti, D., and C. A. Collingwood. 1987. A provisional list of the Balkan ants (Hym., Formicidae) and a key to the worker caste. I. Synonymic list. Mitt. Schweiz. Entomol. Ges. 60:51–62.

Akre, R. D. 1968. The behavior of *Euxenister* and *Pulvinister*, histerid beetles associated with army ants (Coleoptera: Histeridae; Hymenoptera: Formicidae: Dorylinae). Pan-Pac. Entomol. 44:87–101.

Akre, R. D., and C. W. Rettenmeyer. 1966. Behavior of Staphylinidae associated with army ants (Formicidae: Ecitonini). J. Kans. Entomol. Soc. 39:745–782.

Akre, R. D., and C. W. Rettenmeyer. 1968. Trail-following by guests of army ants (Hymenoptera: Formicidae: Ecitonini). J. Kans. Entomol. Soc. 41:165–174.

Akre, R. D., and R. L. Torgerson. 1968. The behavior of *Diploeciton nevermanni*, a staphylinid beetle associated with army ants. Psyche 75:211–215.

Akre, R. D., and R. L. Torgerson. 1969. Behavior of *Vatesus* beetles associated with army ants (Coleoptera: Staphylinidae). Pan-Pac. Entomol. 45:269–281.

Alibert, H. 1951. Les insectes vivant sur les cacaoyers en Afrique occidentale. Mem. Inst. Fr. Agric. Noire 15:1–174.

Arnold, G. 1915. A monograph of the Formicidae of South Africa. Ann. S. Afr. Mus. 14:1–756.

Arnoldi, K. V. 1968. Wichtige Ergänzungen zur Myrmecofauna (Hymenoptera, Formicidae) der USSR, mit einigen Neubeschreibungen. Zool. Zh. 47:1800–1822.

Bagneres, A.-G., J. Billen, and E. D. Morgan. 1991. Volatile secretion of Dufour gland of workers of an army ant, *Dorylus* (*Anomma*) *molestus*. J. Chem. Ecol. 17:1633–1639.

Baker, H. G. 1970. Evolution in the tropics. Biotropica 2: 101–111.

Baldridge, R. S., C. W. Rettenmeyer, and J. F. Watkins II. 1980. Seasonal, nocturnal and diurnal flight periodicities of Nearctic army ant males (Hymenoptera: Formicidae). J. Kans. Entomol. Soc. 53:189–204.

Barlow, E. 1899. Notes on insect pests from the Entomological Section, Indian Museum. Indian Mus. Notes (Calcutta) 4:180–221.

Baroni Urbani, C. 1977. Materiali per una revisione della sottofamiglia Leptanillinae Emery (Hymenoptera: Formicidae). Entomol. Bras. 2:427–488.

Baroni Urbani, C. 1989. Phylogeny and behavioural evolution in ants, with a discussion of the role of behaviour in evolutionary processes. Ethol. Ecol. Evol. 1:137–168.

Baroni Urbani, C., B. Bolton, and P. S. Ward. 1992. The internal phylogeny of ants (Hymenoptera: Formicidae). Syst. Entomol. 17:301–329.

Barr, D., and W. H. Gotwald, Jr. 1982. Phenetic affinities of males of the army ant genus *Dorylus* (Hymenoptera: Formicidae: Dorylinae). Can. J. Zool. 60: 2652–2658.

Barr, D., J. van Boven, and W. H. Gotwald, Jr. 1985. Phenetic studies of African army ant queens of the genus *Dorylus* (Hymenoptera: Formicidae). Syst. Entomol. 10:1–10.

Bartholomew, G. A., J. R. B. Lighton, and D. H. Feener, Jr. 1988. Energetics of trail running, load carriage, and emigration in the column-raiding army ant *Eciton hamatum*. Physiol. Zool. 61:57–68.

Bates, H. W. 1863. The Naturalist on the River Amazons. J. M. Dent & Sons, London. Reprint, No. 446 of Everyman's Library, 1910.

Beattie, A. J., C. L. Turnbull, T. Hough, and R. B. Knox. 1986. Antibiotic production: a possible function for the metapleural glands of ants (Hymenoptera: Formicidae). Ann. Entomol. Soc. Am. 79:448–450.

Beckers, R., S. Goss, J. L. Deneubourg, and J. M. Pasteels. 1989. Colony size, communication and ant foraging strategy. Psyche 96:239–256.

Beebe, W. 1919. The home town of the army ants. Atl. Mon. 124:454–464.

Beebe, W. 1921. Edge of the Jungle. Henry Holt and Company, New York.

Bell, R. H. V. 1971. A grazing ecosystem in the Serengeti. Sci. Am. 225(1):86–93.

Belt, T. 1874. The Naturalist in Nicaragua. John Murray, London.

Bequaert, J. 1913. Notes biologiques sur quelques fourmis et termites du Congo Belge. Rev. Zool. Afr. 2:396–431.

Bequaert, J. 1922. The predaceous enemies of ants. Bull. Am. Mus. Nat. Hist. 45:271–331.

Bernstein, S., and R. A. Bernstein. 1969. Relationships between foraging efficiency and the size of the head and component brain and sensory structures in the red wood ant. Brain Res. 16:85–104.

Billen, J. P. J. 1985. Comparative ultrastructure of the poison and Dufour glands in Old and New World army ants (Hymenoptera, Formicidae). Actes Collog. Insectes Soc. 2:17–26.

Billen, J. P. J. 1987. New structural aspects of the Dufour's and venom glands in social insects. Naturwissenschaften 74:340–341.

Billen, J. 1992. Origin of the trail pheromone in Ecitoninae: a behavioural and morphological examination. Pages 203–209 in J. Billen, ed., Biology and Evolution of Social Insects. Leuven University Press, Leuven, Belgium.

Billen, J. P. J., and W. H. Gotwald, Jr. 1988. The crenellate lining of the Dufour gland in the genus *Aenictus:* a new character for interpreting the phylogeny of Old World army ants (Hymenoptera, Formicidae, Dorylinae). Zool. Scr. 17:293–295.

Billen, J. P. J., and J. K. A. van Boven. 1987. The metapleural gland in Old World army ants: a morphological and ultrastructural description (Hymenoptera, Formicidae). Rev. Zool. Afr. 101:31–41.

Blackwell, M., R. A. Samson, and J. W. Kimbrough. 1980. *Termitariopsis cavernosa,* gen. and sp. nov., a sporodochial fungus ectoparasitic on ants. Mycotaxon 12:97–104.

Blum, M. S. 1974. Pheromonal sociality in the Hymenoptera. Pages 222–249 in M. C. Birch, ed., Pheromones. American Elsevier, New York.

Blum, M. S. 1987. The bases and evolutionary significance of recognitive olfactory acuity in insect societies. Pages 277–293 in J. M. Pasteels and J. L. Deneubourg, eds., From Individual to Collective Behavior in Social Insects. Les Treilles Workshop. Birkhauser Verlag, Basel, Switzerland.

Blum, M. S., J. B. Byrd, J. R. Travis, J. F. Watkins II, and F. R. Gehlbach. 1971. Chemistry of the cloacal sac secretion of the blind snake *Leptotyphlops dulcis.* Comp. Biochem. Physiol. B 38:103–107.

Blum, M. S., and C. A. Portocarrero. 1964. Chemical releasers of social behavior. IV. The hindgut as the source of the odor trail pheromone in the Neotropical army ant genus *Eciton.* Ann. Entomol. Soc. Am. 57:793–794.

Bodot, P. 1961. La destruction des termitières de *Bellicositermes natalensis* Hav., par une fourmi, *Dorylus (Typhlopone) dentifrons* Wasmann. C. R. Hebd. Séances Acad. Sci. 253:3053–3054.

Bodot, P. 1967. Etude écologique des termites des savanes de basse Côte-d'Ivoire. Insectes Soc. 14:229–258.

Bolton, B. 1990a. Abdominal characters and status of the cerapachyine ants (Hymenoptera, Formicidae). J. Nat. Hist. 24:53–68.

Bolton, B. 1990b. The higher classification of the ant subfamily Leptanillinae (Hymenoptera: Formicidae). Syst. Entomol. 15:267–282.

Bolton, B. 1990c. Army ants reassessed: the phylogeny and classification of the doryline section (Hymenoptera, Formicidae). J. Nat. Hist. 24:1339–1364.

Borgmeier, T. 1931. Um cypselideo ecitophio de Goyaz (Dipt. Cypselidae). Rev. Entomol. (Rio de J.) 1:30–37.

Borgmeier, T. 1936. *Ecitotropis,* um novo gênero myrmecophilo da família Staphylinidae (Col.) de Goyaz. Rev. Entomol. 6(2):296–299.

Borgmeier, T. 1939. Sôbre algums Diapriideos myrmecophilos, principalmente do Brasil (Hym. Diapriidae). Rev. Entomol. (Rio de J.) 10:530–545.

Borgmeier, T. 1953. Vorarbeiten zu einer Revision der neotropischen Wanderameisen. Stud. Entomol. 2:1–51.

Borgmeier, T. 1955. Die Wanderameisen der neotropischen Region (Hym. Formicidae). Stud. Entomol. 3:1–717.

Borgmeier, T. 1958. Nachtraege zu meiner Monographie der neotropischen Wanderameisen (Hym. Formicidae). Stud. Entomol. 1(1–2):197–208.

Borgmeier, T. 1963. New or little known *Coniceromyia,* and some other Neo-tropical or Paleotropical Phoridae (Diptera). Stud. Entomol. 6:449–480.

Brauns, J. 1901. Über die Lebenweise von *Dorylus* und *Aenictus.* Z. Syst. Hymenopterol. Dipterol. 1:14–17.

Brosset, A. 1988. Le peuplement de mammifères insectivores des forêts du nord-est du Gabon. Rev. Ecol. (Terre Vie) 43:23–46.

Brothers, D. J. 1975. Phylogeny and classification of the aculeate Hymenoptera, with special reference to the Mutillidae. Univ. Kans. Sci. Bull. 50(11):483–648.

Brown, C. A., J. F. Watkins II, and D. W. Eldridge. 1979. Repression of bacteria and fungi by the army ant secretion: skatole. J. Kans. Entomol. Soc. 52:119–122.

Brown, W. L., Jr. 1960. The release of alarm and attack behavior in some New World army ants. Psyche 66:25–27.

Brown, W. L., Jr. 1968. An hypothesis concerning the function of the metapleural glands in ants. Am. Nat. 102:188–191.

Brown, W. L., Jr. 1975. Contributions toward a reclassification of the Formicidae. V. Ponerinae, tribes Platythyreini, Cerapachyini, Cylindromyrmecini, Acanthostichini, and Aenictogitini. Search Agric. 5:1–116.

Brown, W. L., Jr., and W. L. Nutting. 1950. Wing venation and the phylogeny of the Formicidae (Hymenoptera). Trans. Am. Entomol. Soc. 75:113–132.

Bruch, C. 1923. Estudios mirmecológicos con la descripción de nuevas especies de dipteros ("Phoridae") por los RR. pp. H. Schmitz y Th. Borgmeier y de una araña ("Gonyleptidae") por el Doctor Meilo-Leitao. Rev. Mus. La Plata 27:172–220.

Brues, C. T. 1902. New and little-known guests of the Texan legionary ants. Am. Nat. 36:365–378.

Brues, C. T. 1904. On the relations of certain myrmecophiles to their host ants. Psyche (February) 21–22.

Bulmer, M. G. 1983. Sex ratio theory in social insects with swarming. J. Theor. Biol. 100:329–339.

Burgeon, L. 1924a. Les fourmis "siafu" du Congo. Rev. Zool. Afr. 12:63–65.

Burgeon, L. 1924b. Quelques observations sur les *Dorylus.* Rev. Zool. Afr. 12:429–436.

Burton, J. L., and N. R. Franks. 1985. The foraging ecology of the army ant *Eciton rapax:* an ergonomic enigma? Ecol. Entomol. 10:131–141.

Buschinger, A., C. Peeters, and R. H. Crozier. 1989. Life-pattern studies on an Australian *Sphinctomyrmex* (Formicidae: Ponerinae: Cerapachyini): functional polygyny, brood periodicity and raiding behavior. Psyche 96:287–300.

Byatt, A. S. 1992. Angels and Insects. Two Novellas. Random House, New York.

Caetano, F. H. 1991. Ocorrência de formigas correição, *Labidus praedator* (Hymenoptera Ecitoninae) atacando o gorgulho do milho (*Sitophylus* sp.). An. Soc. Entomol. Bras. 20:453–455.

Campione, B. M., J. A. Novak, and W. H. Gotwald, Jr. 1983. Taxonomy and morphology of the West African army ant, *Aenictus asantei* n.sp. (Hymenoptera: Formicidae). Ann. Entomol. Soc. Am. 76:873–883.

Cardinall, A. W. 1927. In Ashanti and Beyond. Seeley Service and Company, London. Reprint, Negro Universities Press, Westport, Conn., 1970.

Carlin, N. F., and B. Hölldobler. 1983. Nestmate and kin recognition in interspecific mixed colonies of ants. Science 222:1027–1029.

Carpenter, G. D. H. 1920. A Naturalist on Lake Victoria with an Account of Sleeping Sickness and the Tse-tse Fly. T. Fisher Unwin, London.

Carroll, C. R., and D. H. Janzen. 1973. Ecology of foraging by ants. Annu. Rev. Ecol. Syst. 4:231–257.

Chadab, R. 1979. Early warning cues for social wasps attacked by army ants. Psyche 86:115–123.

Chadab, R., and C. W. Rettenmeyer. 1975. Mass recruitment by army ants. Science 188:1124–1125.

Chadab-Crepet, R., and C. W. Rettenmeyer. 1982. Comparative behavior of social wasps when attacked by army ants or other predators and parasites. Pages 270–274 in M. D. Breed, C. D. Michener, and H. E. Evans, eds., The Biology of Social Insects. Proceedings of the Ninth Congress of the International Union for the Study of Social Insects, Boulder, Colo., August 1982. Westview Press, Boulder.

Chapin, J. P. 1932. The birds of the Belgian Congo. Part 1. Bull. Am. Mus. Nat. Hist. 65:1–756.

Chapman, F. M. 1929. My Tropical Air Castle. D. Appleton and Company, New York.

Chapman, J. W. 1964. Studies on the ecology of the army ants of the Philippines genus *Aenictus* Shuckard (Hymenoptera: Formicidae). Philipp. J. Sci. 93:551–595.

Coates-Estrada, R., and A. Estrada. 1989. Avian attendance and foraging at army-ant swarms in the tropical rain forest of Los Tuxtlas, Veracruz, Mexico. J. Trop. Ecol. 5:281–292.

Cohic, F. 1948. Observations morphologiques et écologiques sur *Dorylus (Anomma) nigricans* Illiger (Hymenoptera Dorylidae). Rev. Fr. Entomol. 14:229–275.

Collingwood, C. A. 1978. A provisional list of Iberian Formicidae with a key to the worker caste (Hym. Aculeata). Eos 52:65–96.

Connell, J. H. 1978. Diversity in tropical rain forests and coral reefs. Science 199:1302–1310.

Coody, C. J., and J. F. Watkins II. 1986. The correlation of eye size with circadian flight periodicity of Nearctic army ant males of the genus *Neivamyrmex* (Hymenoptera; Formicidae; Ecitoninae). Tex. J. Sci. 38:3–7.

Cooke, H. B. S. 1972. The fossil mammal fauna of Africa. Pages 89–139 in A. Keast, F. C. Erk, and B. Glass, eds., Evolution, Mammals and Southern Continents. State University of New York Press, Albany.

Craig, R. 1980. Sex investment ratios in social Hymenoptera. Am. Nat. 116:311–323.

Crawley, B. A., and E. Jacobson. 1924. Ants of Sumatra. With biological notes. Ann. Mag. Nat. Hist. 13:380–409.

Crosland, M. W. J., and R. H. Crozier. 1986. *Myrmecia pilosula*, an ant with only one pair of chromosomes. Science 231:1278.

Curio, E. 1976. The Ethology of Predation. Springer-Verlag, Berlin.

Darlington, J. P. E. C. 1985. Attacks of doryline ants and termite nest defences (Hymenoptera; Formicidae; Isoptera; Termitidae). Sociobiology 11:189–200.

Darwin, C. 1872. The Origin of Species by Means of Natural Selection, or the Preservation of Favored Races in the Struggle for Life. 6th ed. Modern Library, New York. Reprint, no date.

DeBach, P. 1974. Biological Control by Natural Enemies. Cambridge University Press, London.

Degallier, N. 1983. Etude des genres *Coelocraera* Marseul et *Coproxenus* Lewis (Coleoptera, Histeridae). Sociobiology 8:137–153.

Dejean, A. 1991. Adaptation d'*Oecophylla longinoda* (Formicidae–Formicinae) aux variations spatio-temorelles de la densité de proies. Entomophaga 36: 29–54.

Delabie, J. C. 1990. The ant problems of cocoa farms in Brazil. Pages 555–569 in R. K. Vander Meer, K. Jaffe, and A. Cedeno, eds., Applied Myrmecology: A World Perspective. Westview Press, Boulder, Colo.

Delamare Deboutteville, C. 1948. Recherches sur les collemboles termitophiles et myrmécophiles (écologie, ethologie, systématique). Arch. Zool. Exp. Gen. 85:261–425.

Deneubourg, J. L., S. Goss, N. Franks, and J. M. Pasteels. 1989. The blind leading the blind: modeling chemically mediated army ant raid patterns. J. Insect Behav. 2:719–725.

Denslow, J. S. 1980. Gap partitioning among tropical rainforest trees. Biotropica 12(suppl.):47–55.

Dietz, R. S., and J. C. Holden. 1970. Reconstruction of Pangaea: breakup and dispersion of continents, Permian to present. J. Geophys. Res. 75:4939–4956.

Dinesen, I. 1937. Out of Africa. Random House, New York. Reprint, Time Inc., New York, 1963.

Disney, R. H., and D. H. Kistner. 1989. Neotropical Phoridae from army ant colonies, including two new species (Diptera; Hymenoptera, Formicidae). Sociobiology 16:149–174.

Disney, R. H. L., and D. H. Kistner. 1991. A new genus of Australasian/Oriental Phoridae associated with driver ants (Diptera; Hymenoptera, Formicidae). Sociobiology 18:269–281.

Donisthorpe, H. S. J. K. 1939. On the occurrence of dealated males in the genus *Dorylus* Fab. (Hym. Formicidae). Proc. R. Entomol. Soc. Lond. 14:79–81.

Droual, R. 1984. Anti-predator behaviour in the ant *Pheidole desertorum:* the importance of multiple nests. Anim. Behav. 32:1054–1058.

Drummond, B. A. 1976. Butterflies associated with an army ant swarm raid in Honduras. J. Lepid. Soc. 30:237–238.

DuBois, M. B. 1988. Distribution of army ants (Hymenoptera: Formicidae) in Illinois. Entomol. News 99:157–160.

Du Chaillu, P. B. 1861. Explorations and Adventures in Equatorial Africa. Harper & Bros., New York. Reprint, Negro Universities Press, Westport, Conn., 1969.

Dutt, G. R. 1912. Life histories of Indian insects. Mem. Dep. Agric., India Entomol. (Ser. 4):183–267.

Dybas, H. S. 1962. Myrmecophiles of the beetle family Limulodidae. Proc. North Cent. Branch Entomol. Soc. Am. 17:15–16.

Eickwort, G. C. 1990. Associations of mites with social insects. Annu. Rev. Entomol. 35:469–488.

Eisner, T. 1957. A comparative morphological study of the proventriculus of ants (Hymenoptera: Formicidae). Bull. Mus. Comp. Zool. 116:437–490.

Eisner, T. 1971. Chemical defense against predation in arthropods. Pages 157–217 in E. Sondheimer and J. B. Simeone, eds., Chemical Ecology. Academic Press, New York.

Eisner, T., J. C. Meinwald, D. W. Alsop, and J. E. Carrel. 1968. Defense mechanisms of arthropods. XXI. Formic acid and n-nonyl-acetate in the defensive spray of two species of *Helluomorphoides*. Ann. Entomol. Soc. Am. 61:610–613.

Ellison, G. T. H. 1988. *Dorylus helvolus* L. (Formicidae: Dorylinae) prey on *Saccostomus campestris* (Rodentia: Cricetidae). J. Entomol. Soc. S. Afr. 51:296.

Elzinga, R. J. 1978. Holdfast mechanisms in certain uropodine mites (Acarina: Uropodina). Ann. Entomol. Soc. Am. 71:896–900.

Elzinga, R. J. 1981. The generic status of six new species of *Trichocylliba* (Acari: Uropodina). Acarologia 23:3–18.

Elzinga, R. J. 1989. *Habeogula cauda* (Acari: Uropodina), a new genus and species of mite from the army ant *Labidus praedator* (F. Smith). Acarologia 30: 341–344.

Elzinga, R. J. 1991. Two new elongate species of *Planodiscus* (Acari: Uropodina) with a key to known species. Acarologia 32:111–114.

Elzinga, R. J. 1993. Larvamimidae, a new family of mites (Acari: Dermanyssoidea) associated with army ants. Acarologia 34:95–103.

Elzinga, R. J., and C. W. Rettenmeyer. 1966. A neotype and new species of *Planodiscus* (Acarina: Uropodina) found on doryline ants. Acarologia 8:191–199.

Elzinga, R. J., and C. W. Rettenmeyer. 1970. Five new species of *Planodiscus* (Acarina: Uropodina) found on doryline ants. Acarologia 12:59–70.

Elzinga, R. J., and C. W. Rettenmeyer. 1974. Seven new species of *Circocylliba* (Acarina: Uropodina) found on army ants. Acarologia 16:595–611.

Emery, C. 1895. Die Gattung *Dorylus* Fab. und die systematische Eintheilung der Formiciden. Zool. Jahrb. Abt. Syst. 8:685–778.

Emery, C. 1910. Subfam. Dorylinae. Genera insectorum, fasc. 102. V. Verteneuil & L. Desmet, Brussels.

Evans, H. E. 1964. A synopsis of the American Bethylidae (Hymenoptera, Aculeata). Bull. Mus. Comp. Zool. 132:1–222.

Fabricius, J. C. 1793. Entomologia systematica. Hafniae, Proft.

Fage, L. 1938. Quelques arachnides provenant de fourmilières ou de termitières du Costa Rica. Bull. Mus. Natl. Hist. Nat. 10:369–376.

Feener, D. H., Jr., J. R. B. Lighton, and G. A. Bartholomew. 1988. Curvilinear allometry, energetics and foraging ecology: a comparison of leaf-cutting ants and army ants. Funct. Ecol. 2:509–520.

Fisher, R. A. 1958. The Genetical Theory of Natural Selection. Dover, New York.

Flanders, S. E. 1976. Revision of a hypothetical explanation for the occasional replacement of a unisexual with a bisexual brood in colonies of the army ant *Eciton*. Bull. Entomol. Soc. Am. 22:133–134.

Forbes, J. 1958. The male reproductive system of the army ant, *Eciton hamatum* Fabricius. Proc. 10th Int. Congr. Entomol. 1:593–596.

Forbes, J., and D. Do-Van-Quy. 1965. The anatomy and histology of the male reproductive system of the legionary ant, *Neivamyrmex harrisi* (Haldeman) (Hymenoptera: Formicidae). J. N.Y. Entomol. Soc. 73:95–111.

Ford, F. C., and J. Forbes. 1980. Anatomy of the male reproductive systems of the adults and pupae of two doryline ants, *Dorylus (Anomma) wilverthi* Emery and *D. (A.) nigricans* Illiger. J. N.Y. Entomol. Soc. 88:133–142.

Ford, F. C., and J. Forbes. 1983. Histology of the male reproductive systems in the adults and pupae of two doryline ants, *Dorylus (Anomma) wilverthi* Emery and *D. (A.) nigricans* Illiger (Hymenoptera: Formicidae). J. N.Y. Entomol. Soc. 91:355–376.

Forel, A. 1896. Ants' nests. Smithson. Rep. 1894:479–505.

Fowler, H. G. 1977. Field responses of *Acromyrmex crassispinus* (Forel) to aggression by *Atta sexdens* (Linn.) and predation by *Labidus praedator* (Fr. Smith) (Hymenoptera: Formicidae). Aggressive Behav. 3:385–391.

Fowler, H. G. 1979. Notes on *Labidus praedator* (Fr. Smith) in Paraguay (Hymenoptera: Formicidae: Dorylinae: Ecitonini). J. Nat. Hist. 13:3–10.

Fowler, H. G. 1984. Recruitment, group retrieval and major worker behavior in *Pheidole oxyops* Forel (Hymenoptera: Formicidae). Rev. Brasil. Biol. 44:21–24.

Franks, N. R. 1982a. A new method for censusing animal populations: the number of *Eciton burchelli* army ant colonies on Barro Colorado Island, Panama. Oecologia 52:266–268.

Franks, N. R. 1982b. Social insects in the aftermath of swarm raids of the army ant *Eciton burchelli*. Pages 275–279 in M. D. Breed, C. D. Michener, and H. E. Evans, eds., The Biology of Social Insects: Proceedings of the Ninth Congress of the International Union for the Study of Social Insects, Boulder, Colo., August 1982. Westview Press, Boulder.

Franks, N. R. 1982c. Ecology and population regulation in the army ant *Eciton burchelli*. Pages 389–395 in E. G. Leigh, Jr., A. S. Rand, and D. M. Windsor, eds., The Ecology of a Tropical Forest: Seasonal Rhythms and Long-Term Changes. Smithsonian Institution Press, Washington, D.C.

Franks, N. R. 1985. Reproduction, foraging efficiency and worker polymorphism in army ants. Pages 9–107 in B. Hölldobler and M. Lindauer, eds., Experimental Behavioral Ecology and Sociobiology. G. Fischer Verlag, Stuttgart.

Franks, N. R. 1986. Teams in social insects: group retrieval of prey by army ants (*Eciton burchelli,* Hymenoptera: Formicidae). Behav. Ecol. Sociobiol. 18:425–429.

Franks, N. R. 1989a. Army ants: a collective intelligence. Am. Sci. 77(2):138–145.

Franks, N. R. 1989b. Thermoregulation in army ant bivouacs. Physiol. Entomol. 14:397–404.

Franks, N. R., and W. H. Bossert. 1983. The influence of swarm raiding army ants on the patchiness and diversity of a tropical leaf litter ant community. Pages 151–163 in S. L. Sutton, T. C. Whitmore, and A. C. Chadwick, eds., Tropical Rain Forest: Ecology and Management. Blackwell, Oxford.

Franks, N. R., and C. R. Fletcher. 1983. Spatial patterns in army ant foraging and migration: *Eciton burchelli* on Barro Colorado Island, Panama. Behav. Ecol. Sociobiol. 12:261–270.

Franks, N. R., and B. Hölldobler. 1987. Sexual competition during colony reproduction in army ants. Biol. J. Linn. Soc. 30:229–243.

Garcia Márquez, Gabriel. 1970. One Hundred Years of Solitude. Harper & Row, New York.

Gehlbach, F. R., J. F. Watkins II, and J. C. Kroll. 1971. Pheromone trailfollowing studies of typhlopid, leptotyphlopid, and colubrid snakes. Behaviour 40:282–294.

Gehlbach, F. R., J. F. Watkins II, and H. W. Reno. 1968. Blind snake defensive behavior elicited by ant attacks. BioScience 18:784–785.

Gerstaecker, A. 1863. Ueber ein merkwürdiges neues Hymenopteron aus der Abtheilung der Aculeata. Stett. Entomol. Z. 24:76–93.

Ghosh, C. C. 1936. Bee-keeping. Miscellaneous Bulletin no. 6. Imperial Council of Agricultural Research, Delhi. 91 pp.

Gobin, B., J. Billen, N. J. Oldham, and E. D. Morgan. 1993. A novel exocrine gland in the army ant *Aenictus* and its function in trail following. Belg. J. Zool. 123(supp. 1):29.

Goodall, J. 1963. Feeding behaviour of wild chimpanzees. A preliminary report. Symp. Zool. Soc. Lond. 10:39–48.

Gotwald, W. H., Jr. 1969. Comparative morphological studies of the ants, with particular reference to the mouthparts (Hymenoptera: Formicidae). Cornell Univ. Agric. Exp. Stn. Mem. 408:1–150.

Gotwald, W. H., Jr. 1971. Phylogenetic affinities of the ant genus *Cheliomyrmex* (Hymenoptera: Formicidae). J. N.Y. Entomol. Soc. 79:161–173.

Gotwald, W. H., Jr. 1972a. Analogous prey escape mechanisms in a pulmonate mollusk and lepidopterous larvae. J. N.Y. Entomol. Soc. 80:111–113.

Gotwald, W. H., Jr. 1972b. *Oecophylla longinoda*, an ant predator of *Anomma* driver ants (Hymenoptera: Formicidae). Psyche 79:348–356.

Gotwald, W. H., Jr. 1974a. Predatory behavior and food preferences of driver ants in selected African habitats. Ann. Entomol. Soc. Am. 67:877–886.

Gotwald, W. H., Jr. 1974b. Foraging behavior of *Anomma* driver ants in Ghana cocoa farms (Hymenoptera: Formicidae). Bull. Inst. Fondam. Afr. Noire, Ser. A 36:705–713.

Gotwald, W. H., Jr. 1976. Behavioral observations on African army ants of the genus *Aenictus* (Hymenoptera: Formicidae). Biotropica 8:59–65.

Gotwald, W. H., Jr. 1977. The origins and dispersal of army ants of the subfamily Dorylinae. Pages 126–127 in H. H. W. Velthuis and J. T. Wiebes, eds., Proceedings of the Eighth International Congress of the International Union for the Study of Social Insects. Centre for Agricultural Publishing and Documentation, Wageningen, The Netherlands.

Gotwald, W. H., Jr. 1978a. Emigration behavior of the East African driver ant, *Dorylus* (*Anomma*) *molesta* Gerstaecker (Hymenoptera: Formicidae: Dorylinae). J. N.Y. Entomol. Soc. 86:290.

Gotwald, W. H., Jr. 1978b. Trophic ecology and adaptation in tropical Old World ants of the subfamily Dorylinae (Hymenoptera: Formicidae). Biotropica 10:161–169.

Gotwald, W. H., Jr. 1979. Phylogenetic implications of army ant zoogeography (Hymenoptera: Formicidae). Ann. Entomol. Soc. Am. 72:462–467.

Gotwald, W. H., Jr. 1982. Army ants. Pages 157–254 in H. R. Hermann, ed., Social Insects, vol. 4. Academic Press, New York.

Gotwald, W. H., Jr. 1984. Lizard eggs in an arboreal carton nest of Neotropical ant *Azteca* sp.: example of a nest/egg-laying association between vertebrates and aculeate Hymenoptera. Sociobiology 9:9–18.

Gotwald, W. H., Jr. 1984–1985. Death on the march. Rotunda 17(3):37–41.

Gotwald, W. H., Jr. 1985. Reflections on the evolution of army ants (Hymenoptera, Formicidae). Actes Colloq. Insectes Soc. 2:7–16.

Gotwald, W. H., Jr. 1986. The beneficial economic role of ants. Pages 290–313 in S. B. Vinson, ed., Economic Impact and Control of Social Insects. Praeger, New York.

Gotwald, W. H. Jr. 1987. The relationship of form and function in army ant queens. Pages 255–256 in J. Eder and H. Rembold, eds., Chemistry and Biology of Social Insects: Proceedings of the Tenth International Congress of the International Union for the Study of Social Insects. J. Peperny Verlag, Munich.

Gotwald, W. H., Jr. 1988. On becoming an army ant. Pages 227–235 in J. C. Trager, ed., Advances in Myrmecology. E. J. Brill, Leiden.

Gotwald, W. H., Jr. 1991. Terrors of the jungle. Wildl. Conserv. 94(6):64–71.

Gotwald, W. H., Jr. 1994. Army ant guests and associates: a new analysis of functional relationships. Page 345 in A. Lenoir, G. Arnold, and M. Lepage, eds., Les insectes sociaux. 12th International Congress of the International Union for the Study of Social Insects. Université Paris-Nord, Paris.

Gotwald, W. H., Jr., and D. Barr. 1980. Quantitative studies on the major workers of the ant genus *Dorylus* (Hymenoptera: Formicidae: Dorylinae). Ann. Entomol. Soc. Am. 73:231–238.

Gotwald, W. H., Jr., and D. Barr. 1987. Quantitative studies on workers of the Old World army ant genus *Aenictus* (Hymenoptera: Formicidae). Insectes Soc. 34:261–273.

Gotwald, W. H., Jr., and W. L. Brown, Jr. 1966. The ant genus *Simopelta* (Hymenoptera: Formicidae). Psyche 73:261–277.

Gotwald, W. H., Jr., and A. W. Burdette. 1981. Morphology of the male internal reproductive system in army ants: phylogenetic implications (Hymenoptera: Formicidae). Proc. Entomol. Soc. Wash. 83:72–92.

Gotwald, W. H., Jr., and G. R. Cunningham–van Someren. 1976. Taxonomic and behavioral notes on the African ant, *Aenictus eugenii* Emery, with a description of the queen (Hymenoptera: Formicidae). J. N.Y. Entomol. Soc. 84:182–188.

Gotwald, W. H., Jr., and G. R. Cunningham–van Someren. 1990. A year in the life of an Old World army ant colony: spatial patterns in foraging and emigration. Pages 714–715 in G. K. Veeresh, B. Mallik, and C. A. Viraktamath, eds., Social Insects and the Environment: Proceedings of the Eleventh International Congress of the International Union for the Study of Social Insects. Oxford and IBH Publishing Co., New Delhi.

Gotwald, W. H., Jr., and B. M. Kupiec. 1975. Taxonomic implications of doryline worker ant morphology: *Cheliomyrmex morosus* (Hymenoptera: Formicidae). Ann. Entomol. Soc. Am. 68:961–971.

Gotwald, W. H., Jr., and J. M. Leroux. 1980. Taxonomy of the African army ant, *Aenictus decolor* (Mayr), with a description of the queen (Hymenoptera: Formicidae). Proc. Entomol. Soc. Wash. 82:599–608.

Gotwald, W. H., Jr., and R. F. Schaefer, Jr. 1982. Taxonomic implications of doryline worker ant morphology: *Dorylus* subgenus *Anomma* (Hymenoptera: Formicidae). Sociobiology 7:187–204.

Green, E. E. 1903. Notes on *Dorylus orientalis*, West. Indian Mus. Notes (Calcutta) 5:39.

Gudger, E. W. 1925. Stitching wounds with the mandibles of ants and beetles: a minor contribution to the history of surgery. J. Am. Med. Assoc. 84:1861–1864.

Haddow, A. J., H. H. Yarrow, G. A. Lancaster, and P. S. Corbet. 1966. Nocturnal flight cycle in the males of African doryline ants (Hymenoptera: Formicidae). Proc. R. Entomol. Soc. Lond. 41:103–106.

Haemig, P. D. 1989. Brown jays as army ant followers. Condor 91:1008–1009.

Hagan, H. R. 1954a. The reproductive system of the army-ant queen, *Eciton (Eciton)*. Part 1. General anatomy. Am. Mus. Novit. 1663:1–12.

Hagan, H. R. 1954b. The reproductive system of the army-ant queen, *Eciton (Eciton)*. Part 2. Histology. Am. Mus. Novit. 1664:1–17.

Hagan, H. R. 1954c. The reproductive system of the army-ant queen, *Eciton (Eciton)*. Part 3. The oocyte cycle. Am. Mus. Novit. 1665:1–20.

Haley, A. 1976. Roots. Doubleday, Garden City, N.Y.

Hamilton, W. D. 1972. Altruism and related phenomena, mainly in social insects. Annu. Rev. Ecol. Syst. 3:193–232.

Hamilton, W. D. 1975. Gamblers since life began: barnacles, aphids, elms. Q. Rev. Biol. 50:175–180.

Harcourt, A. H., and S. A. Harcourt. 1984. Insectivory by gorillas. Folia Primatol. 43:229–233.

Hashimoto, Y. 1991. Phylogenetic study of the family Formicidae based on the sensillum structures on the antennae and labial palpi (Hymenoptera, Aculeata). Jpn. J. Entomol. 59:125–140.

Hepburn, K. 1987. The Making of *The African Queen*, or How I Went to Africa with Bogart, Bacall and Huston and Almost Lost My Mind. Knopf, New York.

Hermann, H. R., Jr. 1968a. Group raiding in *Termitopone commutata* (Roger) (Hymenoptera: Formicidae). J. Ga. Entomol. Soc. 3:23–24.

Hermann, H. R. 1968b. The hymenopterous poison apparatus. VII. *Simopelta oculata* (Hymenoptera: Formicidae: Ponerinae). J. Ga. Entomol. Soc. 3:163–166.

Hermann, H. R. 1969. The hymenopterous poison apparatus: evolutionary trends in three closely related subfamilies of ants (Hymenoptera: Formicidae). J. Ga. Entomol. Soc. 4:123–141.

Hermann, H. R., and J. T. Chao. 1983. Furcula, a major component of the hymenopterous venom apparatus. Int. J. Insect Morphol. Embryol. 12:321–337.

Herre, E. A., D. M. Windsor, and R. B. Foster. 1986. Nesting associations of wasps and ants on lowland Peruvian ant-plants. Psyche 93:321–330.

Hölldobler, B. 1983. Territorial behavior in the green tree ant (*Oecophylla smaragdina*). Biotropica 15:241–250.

Hölldobler, B. 1984. Evolution of insect communication. Pages 349–377 in T. Lewis, ed., Insect Communication. Symposium of the Royal Entomological Society of London, no. 12. Academic Press, London.

Hölldobler, B., and H. Engel. 1978. Tergal and sternal glands in ants. Psyche 85:285–330.

Hölldobler, B., and H. Engel-Siegel. 1982. Tergal and sternal glands in male ants. Psyche 89:113–132.

Hölldobler, B., and H. Engel-Siegel. 1984. On the metapleural gland of ants. Psyche 91:201–224.

Hölldobler, B., J. M. Palmer, K. Masuko, and W. L. Brown, Jr. 1989. New exocrine glands in the legionary ants of the genus *Leptanilla* (Hymenoptera, Formicidae, Leptanillinae). Zoomorphology 108:255–261.

Hölldobler, B., and E. O. Wilson. 1977. The number of queens: an important trait in ant evolution. Naturwissenschaften 64:8–15.

Hölldobler, B., and E. O. Wilson. 1990. The Ants. Harvard University Press, Cambridge.

Holliday, M. 1904. A study of some ergatogynic ants. Zool. Jahrb. Abt. Syst. 19:293–328.

Hollingsworth, M. J. 1960. Studies on the polymorphic workers of the army ant *Dorylus* (*Anomma*) *nigricans* Illiger. Insectes Soc. 7:17–37.

Hood, W. G., and W. R. Tschinkel. 1990. Desiccation resistance in arboreal and terrestrial ants. Physiol. Entomol. 15:23–35.

Huang, H. T., and P. Yang. 1987. The ancient cultured citrus ant. BioScience 37:665–671.

Huggert, L., and L. Masner. 1983. A review of myrmecophilic-symphilic diapriid wasps in the Holarctic realm, with descriptions of new taxa and a key to genera (Hymenoptera: Proctotrupoidea: Diapriidae). Contrib. Am. Entomol. Inst. 20:63–89.

Hung, A. C. F., H. T. Imai, and M. Kubota. 1972. The chromosomes of nine ant species (Hymenoptera: Formicidae) from Taiwan, Republic of China. Ann. Entomol. Soc. Am. 65:1023–1025.

Hung, A. C. F., and S. B. Vinson. 1975. Notes on the male reproductive system in ants (Hymenoptera: Formicidae). J. N.Y. Entomol. Soc. 83:192–197.

Hunt, J. H. 1983. Foraging and morphology in ants: the role of vertebrate predators as agents of natural selection. Pages 83–101 in P. Jaisson, ed., Social Insects in the Tropics, vol. 2. Université Paris-Nord, Paris.

Huxley, J. S. 1927. Further work on heterogonic growth. Biol. Zentralbl. 47:151–163.

Imai, H. T., C. Baroni Urbani, M. Kubota, G. P. Sharma, M. N. Narasimhanna, B. C. Das, A. K. Sharma, A. Sharma, and G. B. Deodikar. 1984a. Karyological survey of Indian ants. Jpn. J. Genet. 59:1–32.

Imai, H. T., W. L. Brown, Jr., M. Kubota, H. S. Yong, and Y. P. Tho. 1984b. Chromosome observations on tropical ants from western Malaysia. II. Annu. Rep. Natl. Inst. Genet. (Jpn.) 34:66–69.

Jackson, W. B. 1957. Microclimatic patterns in the army ant bivouac. Ecology 38:276–285.

Jacobson, H. R., and D. H. Kistner. 1975a. New species and new records of the Pygostenini. Sociobiology 1:163–200.

Jacobson, H. R., and D. H. Kistner. 1975b. A manual for the identification of the Pygostenini. Sociobiology 1:201–335.

Jacobson, H. R., and D. H. Kistner. 1981. A revision of the genus *Homalodonia* with a description of their behavior and relationships (Coleoptera, Staphylinidae, Zyrasini). Sociobiology 6:185–209.

Jacobson, H. R., and D. H. Kistner. 1983. Revision of some myrmecoid Zyrasini from Africa (Coleoptera: Staphylinidae) with notes on their behavior and relationship. Sociobiology 8:1–50.

Janzen, D. H. 1969. Seed-eaters versus seed size, number, toxicity and dispersal. Evolution 23:1–27.

Janzen, D. H. 1970. Herbivores and the number of tree species in tropical forests. Am. Nat. 104:501–528.

Jeanne, R. L. 1970. Chemical defense of brood by a social wasp. Science 168: 1465–1466.

Jeanne, R. L. 1975. The adaptiveness of social wasp nest architecture. Q. Rev. Biol. 50:267–287.

Jessen, K. 1987. Gastral exocrine glands in ants—functional and systematical aspects. Pages 445–446 in J. Eder and H. Rembold, eds., Chemistry and Biology of Social Insects. J. Peperny Verlag, Munich.

Johnson, R. A. 1954. The behavior of birds attending army ant raids on Barro Colorado Island, Panama Canal Zone. Proc. Linn. Soc. N.Y. 63–65: 41–70.

Kannowski, P. B. 1969. Daily and seasonal periodicities in the nuptial flights of Neotropical ants. I. Dorylinae. Pages 77–83 in Proceedings of the Sixth International Congress of the International Union for the Study of Social Insects. Zoological Institute, University of Bern, Bern, Switzerland.

Karr, J. R. 1976. Within- and between-habitat avian diversity in African and Neotropical lowland habitats. Ecol. Monogr. 46:457–481.

Keegans, S. J., J. Billen, D. Morgan, and O. A. Gökcen. 1993. Volatile glandular secretions of three species of New World army ants, *Eciton burchelli*, *Labidus coecus*, and *Labidus praedator*. J. Chem. Ecol. 19:2705–2719.

Kistner, D. H. 1966a. A revision of the myrmecophilous tribe Deremini (Coleoptera: Staphylinidae). I. The *Dorylopora* complex and their behavior. Ann. Entomol. Soc. Am. 59:341–358.

Kistner, D. H. 1966b. A revision of the African species of the aleocharine tribe Dorylomimini (Coleoptera: Staphylinidae). II. The genera *Dorylomimus*, *Dorylonannus*, *Dorylogaster*, *Dorylobactrus*, and *Mimanomma*, with notes on their behavior. Ann. Entomol. Soc. Am. 59:320–340.

Kistner, D. H. 1976. The natural history of the myrmecophilous tribe Pygostenini (Coleoptera: Staphylinidae). Section 3. Behavior and food habits of the Pygostenini. Sociobiology 2:171–188.

Kistner, D. H. 1979. Social and evolutionary significance of social insect symbionts. Pages 339–413 in H. R. Hermann, ed., Social Insects, vol. 1. Academic Press, New York.

Kistner, D. H. 1982. The social insects' bestiary. Pages 1–244 in H. R. Hermann, ed., Social Insects, vol. 3. Academic Press, New York.

Kistner, D. H., and L. N. Davis. 1989. New species of *Notoxopria* and their behavior, with notes on *Mimopria* (Hymenoptera: Diapriidae). Sociobiology 16:217–238.

Kistner, D. H., and H. R. Jacobson. 1975a. A review of the myrmecophilous Staphylinidae associated with *Aenictus* in Africa and the Orient (Coleoptera; Hymenoptera, Formicidae) with notes on their behavior and glands. Sociobiology 1:21–73.

Kistner, D. H., and H. R. Jacobson. 1975b. An annotated catalog of the Pygostenini. Sociobiology 1:336–371.

Kistner, D. H., and H. R. Jacobson. 1981. A revision of the genera *Creodonia* and *Aulacocephalonia*, with notes on their behavior and relationships (Coleoptera: Staphylinidae). Sociobiology 6:61–100.

Kistner, D. H., and H. R. Jacobson. 1982. A revision of the genera *Trichodonia*, *Gapia*, *Myrmechusina*, and *Myrmechusa* (Coleoptera, Staphylinidae) with an analysis of their relationships and notes on their behavior. Sociobiology 7: 73–128.

Kistner, D. H., and H. R. Jacobson. 1990. Cladistic analysis and taxonomic revision of the ecitophilous tribe Ecitocharini with studies of their behavior and evolution (Coleoptera, Staphylinidae, Aleocharinae). Sociobiology 17: 333–465.

Kruuk, H., and W. A. Sands. 1972. The aardwolf (*Proteles cristatus* Sparrman) 1783 as predator of termites. East Afr. Wildl. J. 10:211–227.

Kugler, C. 1992. Stings of ants of the Leptanillinae (Hymenoptera: Formicidae). Psyche 99:103–115.

Kugler, J. 1986. The Leptanillinae (Hymenoptera: Formicidae) of Israel and a description of a new species from India. Isr. J. Entomol. 20:45–57.

Kukuk, P. F., G. C. Eickwort, M. Raveret-Richter, B. Alexander, R. Gibson, R. A. Morse, and F. Ratnieks. 1989. Importance of the sting in the evolution of sociality in the Hymenoptera. Ann. Entomol. Soc. Am. 82:1–5.

Kutter, H. 1948. Beitrag zur Kenntnis der Leptanillinae (Hym. Formicidae). Eine neue Ameisengattung aus Sud-Indien. Mitt. Schweiz. Entomol. Ges. 21:286–295.

Lamas, G. 1983. Mariposas atraídas por hormigas legionarias en la reserva de Tambopata, Perú. Rev. Soc. Méx. Lepidopterol A. C. 8:49–51.

Lamborn, W. A. 1913–1914. Observations on the driver ants (*Dorylus*) of southern Nigeria. Proc. R. Entomol. Soc. Lond. 5:123–129.

Lamborn, W. A. 1914–1915. Further notes on the driver ants (*Dorylus*) of southern Nigeria. Proc. R. Entomol. Soc. Lond., pp. 5–8.

LaMon, B., and H. Topoff. 1981. Avoiding predation by army ants: defensive behaviors of three ant species of the genus *Camponotus*. Anim. Behav. 29: 1070–1081.

Lappano, E. R. 1958. A morphological study of larval development in polymorphic all-worker broods of the army ant *Eciton burchelli*. Insectes Soc. 5: 31–66.

Leach, W. 1815. Entomology. Pages 646–758 in D. Brewster, ed., The Edinburgh Encyclopaedia. 1st American ed., vol. 9. Joseph and Edward Parker, Philadelphia.

Lenko, K. 1969. An army ant attacking the "guaia" crab in Brazil. Entomol. News 80:6.

Leroux, J. M. 1975. Recherches sur les nids et l'activité prédatrice des dorylines *Anomma nigricans* Illiger (Hym. Formicidae). Mémoire présenté á l'Université Pierre et Marie Curie, Paris, pour l'Diplome d'Etudes Supérieures de Sciences Naturelles.

Leroux, J. M. 1977a. Densité des colonies et observations sur les nids de dorylines *Anomma nigricans* Illiger (Hym. Formicidae) dans la région de Lamto (Côte d'Ivoire). Bull. Soc. Zool. Fr. 102:51–62.

Leroux, J. M. 1977b. Formation et déroulement des raids de chasse d'*Anomma nigricans* Illiger (Hym. Dorylinae) dans une savane de Côte d'Ivoire. Bull. Soc. Zool. Fr. 102:445–458.

Leroux, J. M. 1979a. Possibilités de scissions multiples pour des colonies de dorylines *Anomma nigricans* Illiger (Hyménoptères Formicidae) en Côte d'Ivoire. Insectes Soc. 26:13–17.

Leroux, J. M. 1979b. Sur quelques modalités de disparition des colonies d'*Anomma nigricans* Illiger (Formicidae Dorylinae) dans la région de Lamto (Côte d'Ivoire). Insectes Soc. 26:93–100.

Leroux, J. M. 1982. Ecologie des populations de dorylines *Anomma nigricans* Illiger (Hym. Formicidae) dans la région de Lamto (Côte d'Ivoire). Publications du Laboratoire de Zoologie, no. 22. Ecole Normale Supérieure, Paris.

Leston, D. 1979. Dispersal by male doryline ants in West Africa. Psyche 86:63–77.

Levings, S. C., and N. R. Franks. 1982. Patterns of nest dispersion in a tropical ground ant community. Ecology 63:338–344.

Linden, E. 1992. The last Eden. Time 140(2):62–68.

Lindquist, A. W. 1942. Ants as predators of *Cochliomyia americana* C. & P. J. Econ. Entomol. 35:850–852.

Linné, C. 1764. Insecta & Conchilla. Museum Ludoviciae Ulricae Reginae: Literis & Impensis Direct. Laur. Salvii.

Linné, C. 1767. Systema naturae, 12th ed. Vol. 1, Pars 2, Classis 5, Insecta.

Liu, G. 1939. Some extracts from the history of entomology in China. Psyche 46:23–28.

Long, W. H., Jr. 1902. New species of *Ceratopogon*. Biol. Bull. (Woods Hole) 3:3–14.

Lovejoy, T. E., R. O. Bierregaard, J. M. Rankin, and H. O. R. Schubart. 1983. Ecological dynamics of tropical forest fragments. Pages 377–384 in S. L. Sutton, T. C. Whitmore, and A. C. Chadwick, eds., Tropical Rain Forest: Ecology and Management. Blackwell, Oxford.

Lovejoy, T. E., R. O. Bierregaard, Jr., A. B. Rylands, J. R. Malcolm, C. E. Quintela, L. H. Harper, K. S. Brown, Jr., A. H. Powell, G. V. N. Powell, H. O. R. Schubart, and M. B. Hays. 1986. Edge and other effects of isolation on Amazon forest fragments. Pages 257–285 in M. E. Soulé, ed., Conservation Biology: The Science of Scarcity and Diversity. Sinauer, Sunderland, Mass.

Lovejoy, T. E., J. M. Rankin, R. O. Bierregaard, Jr., K. S. Brown, Jr., L. H. Emmons, and M. E. Vander Voort. 1984. Ecosystem decay of Amazon forest remnants. Pages 295–335 in M. H. Nitecki, ed., Extinctions. University of Chicago Press, Chicago.

Loveridge, A. 1922. Account of an invasion of "siafu" or red driver-ants— *Dorylus* (*Anomma*) *nigricans* Illig. Proc. R. Entomol. Soc. Lond. 5:33–46.

Loveridge, A. 1949. Many Happy Days I've Squandered. Scientific Book Club, London.

Lund, M. 1831. Lettre sur les habitudes de quelques fourmis du Brésil, adressée à M. Audouin. Ann. Sci. Nat. 23:113–138.

Macevicz, S. 1979. Some consequences of Fisher's sex ratio principle for social Hymenoptera that reproduce by colony fission. Am. Nat. 113:363–371.

Mahto, Y. 1991. Varietal susceptibility of *Dorylus orientalis* Westwood (Hymenoptera: Formicidae) in groundnut varieties. J. Entomol. Res. (New Delhi) 15:144–148.

Mahunka, S. 1977a. The examination of myrmecophilous tarsonemid mites based on the investigation of Dr. C. W. Rettenmeyer (Acari). I. Acta Zool. Acad. Sci. Hung. 23:99–132.

Mahunka, S. 1977b. The examination of myrmecophilous tarsonemid mites based on the investigations of Dr. C. W. Rettenmeyer (Acari). II. Acta Zool. Acad. Sci. Hung. 23:341–370.

Mahunka, S. 1978. The examination of myrmecophilous Acaroidea mites based on the investigations of Dr. C. W. Rettenmeyer (Acari: Acaroidea). I. Folia Entomol. Hung. 31:135–166.

Mahunka, S. 1979. The examination of myrmecophilous Acaroidea mites based on the investigations of Dr. C. W. Rettenmeyer (Acari: Acaroidea). II. Acta Zool. Acad. Sci. Hung. 25:311–342.

Majer, J. D. 1974. The use of ants in an integrated control scheme for cocoa. Pages 181–190 in Proceedings of the Fourth Conference of West African Cocoa Entomologists. University of Ghana, Legon.

Majer, J. D. 1976a. The maintenance of the ant mosaic in Ghana cocoa farms. J. Appl. Ecol. 13:123–144.

Majer, J. D. 1976b. The ant mosaic in Ghana cocoa farms: further structural considerations. J. Appl. Ecol. 13:145–155.

Majer, J. D. 1976c. The influence of ants and ant manipulation on the cocoa farm fauna. J. Appl. Ecol. 13:157–175.

Mann, W. M. 1912. Note on a guest of *Eciton hamatum* Fabr. Psyche 19:98–100.

Mann, W. M. 1923. Two serphoid guests of *Eciton* (Hym.). Proc. Entomol. Soc. Wash. 25:181–182.

Mann, W. M. 1925. Guests of *Eciton hamatum* (Fab.) collected by Professor W. M. Wheeler. Psyche 32:166–177.

Maschwitz, U., K. Koob, and H. Schildknecht. 1970. Ein Beitrag zur Funktion der Metapleuraldruse der Ameisen. J. Insect Physiol. 16:387–404.

Maschwitz, U., and P. Schonegge. 1980. Fliegen als Beute und Brutrauber bei Ameisen. Insectes Soc. 27:1–4.

Maschwitz, U., S. Steghaus-Kovac, R. Gaube, and H. Hanel. 1989. A South East Asian ponerine ant of the genus *Leptogenys* (Hym., Form.) with army ant life habits. Behav. Ecol. Sociobiol. 24:305–316.

Masner, L. 1976. Notes on the ecitophilous diapriid genus *Mimopria* Holmgren (Hymenoptera: Proctotrupoidea, Diapriidae). Can. Entomol. 108:123–126.

Masner, L. 1977. A new genus of ecitophilous diapriid wasps from Arizona (Hymenoptera: Proctotrupoidea: Diapriidae). Can. Entomol. 109:33–36.

Masuko, K. 1987. *Leptanilla japonica:* the first bionomic information on the enigmatic ant subfamily Leptanillinae. Pages 597–598 in J. Eder and H. Rembold, eds., Chemistry and Biology of Social Insects. Proceedings of the Tenth International Congress of the International Union for the Study of Social Insects. J. Peperny Verlag, Munich.

Masuko, K. 1989. Larval hemolymph feeding in the ant *Leptanilla japonica* by use of a specialized duct organ, the "larval hemolymph tap" (Hymenoptera: Formicidae). Behav. Ecol. Sociobiol. 24:127–132.

Masuko, K. 1990. Behavior and ecology of the enigmatic ant *Leptanilla japonica* Baroni Urbani (Hymenoptera: Formicidae: Leptanillinae). Insectes Soc. 37: 31–57.

Mayr, E. 1982. The Growth of Biological Thought: Diversity, Evolution, and Inheritance. Harvard University Press, Cambridge.

McDonald, P., and H. Topoff. 1986. The development of defensive behavior against predation by army ants. Dev. Psychobiol. 19:351–367.

McGrew, W. C. 1974. Tool use by wild chimpanzees in feeding upon driver ants. J. Hum. Evol. 3:501–508.

Menon, M. G. R., and M. L. Srivastava. 1976. *Dorylus orientalis* (Westwood) as a pest of mango fruits (Hymenoptera: Formicidae). Entomol. Newsl. 6:2.

Mirenda, J. T., D. G. Eakins, K. Gravelle, and H. Topoff. 1980. Predatory behavior and prey selection by army ants in a desert-grassland habitat. Behav. Ecol. Sociobiol. 7:119–127.

Mirenda, J. T., and H. Topoff. 1980. Nomadic behavior of army ants in a desert-grassland habitat. Behav. Ecol. Sociobiol. 7:129–135.

Moffett, M. W. 1984. Swarm raiding in a myrmicine ant. Naturwissenschaften 71:588–589.

Moffett, M. W. 1987. Division of labor and diet in the extremely polymorphic ant *Pheidologeton diversus*. Natl. Geogr. Res. 3:282–304.

Moffett, M. W. 1988a. Cooperative food transport in an Asiatic ant. Natl. Geogr. Res. 4:386–394.

Moffett, M. W. 1988b. Nesting, emigrations, and colony foundation in two group-hunting myrmicine ants. Pages 355–370 in J. C. Trager, ed., Advances in Myrmecology. E. J. Brill, Leiden.

Moore, W. 1913. The maize stalk borer and its control. Agric. J. Union S. Afr. 5:419–428.

Mukerjee, D. 1926. Digestive and reproductive systems of the male ant *Dorylus labiatus* Shuck. J. Asiat. Soc. Bengal 22:87–91.

Mukerjee, D. 1933. On the anatomy of the worker of the ant *Dorylus (Alaopone) orientalis* Westw. Zool. Anz. 105:97–105.

Murray-Brown, J. 1973. Kenyatta. E. P. Dutton, New York.

Myers, N. 1983. Conversion rates in tropical moist forests. Pages 289–300 in F. B. Golley, ed., Tropical Rain Forest Ecosystems: Structure and Function. Ecosystems of the World 14A. Elsevier, Amsterdam.

O'Donnell, S., and R. L. Jeanne. 1990. Notes on an army ant (*Eciton burchelli*) raid on a social wasp colony (*Agelaia yepocapa*) in Costa Rica. J. Trop. Ecol. 6:507–509.

Oniki, Y. 1972. Studies of the guild of ant-following birds at Belem, Brazil. Acta Amazonica 2:59–79.

Orians, G. H., and N. E. Pearson. 1979. On the theory of central place foraging. Pages 155–177 in D. J. Horn, G. R. Stairs, and R. D. Mitchell, eds., Analysis of Ecological Systems. Ohio State University Press, Columbus.

Otis, G. W., E. Santana C., D. L. Crawford, and M. L. Higgins. 1986. The effect of foraging army ants on leaf-litter arthropods. Biotropica 18:56–61.

Park, O. 1933. Ecological study of the ptiliid myrmecocole, *Limulodes paradoxus* Matthews. Ann. Entomol. Soc. Am. 26:255–261.

Patrizi, S. 1946. Stomach contents of a female ant-eater from Nairobi. J. East Afr. Nat. Hist. Soc. 19:67–68.

Paulian, R. 1948. Observations sur les Coléoptères commensaux d'*Anomma nigricans* en Côte d'Ivoire. Ann. Sci. Nat. Zool. Biol. Anim. 10:79–102.

Perkins, G. A. 1869. The drivers. Am. Nat. 3:360–364.

Petralia, R. S., and S. B. Vinson. 1979. Comparative anatomy of the ventral region of ant larvae, and its relation to feeding behavior. Psyche 86:375–394.

Pisarski, B. 1967. Fourmis (Hymenoptera: Formicidae) d'Afghanistan récoltées par M. Dr. K. Lindberg. Ann. Zool. Polska Akad. Nauk. 24:375–425.

Plsek, R. W., J. C. Kroll, and J. F. Watkins II. 1969. Observations of carabid beetles, *Helluomorphoides texanus*, in columns of army ants and laboratory experiments on their behavior. J. Kans. Entomol. Soc. 42:452–456.

Prins, A. J., H. G. Roberton, and A. Prins. 1990. Pest ants in urban and agricultural areas of southern Africa. Pages 25–33 in R. K. Vander Meer, K. Jaffe, and A. Cedeno, eds., Applied Myrmecology: A World Perspective. Westview Press, Boulder, Colo.

Pullen, B. E. 1963. Termitophagy, myrmecophagy, and the evolution of the Dorylinae (Hymenoptera, Formicidae). Stud. Entomol. 6:405–414.

Raignier, A. 1959. Het ontstaan van kolonies en koninginner bij de Afrikaanse trekmieren (Formicidae, Dorylinae, Dorylini). Meded. K. Acad. Wet. Lett. Schone Kunsten Belg. Kl. Wet. 21:3–24.

Raignier, A. 1972. Sur l'origine des nouvelles sociétés des fourmis voyageuses africaines. Insectes Soc. 19:153–170.

Raignier, A., and J. K. A. van Boven. 1955. Etude taxonomique, biologique et biométrique des *Dorylus* du sous-genre *Anomma* (Hymenoptera Formicidae). Ann. Mus. R. Congo Belg. 2:1–359.

Rajagopal, D., B. Krishnappa, and J. W. M. Logan. 1990. Army ant, *Dorylus orientalis*, a pest of groundnut. Pages 126–128 in G. K. Veeresh, A. R. V. Kumar, and T. Shivashankar, eds., Social Insects: An Indian Perspective. Proceedings of the First National Symposium on Social Insects. International Union for the Study of Social Insects, Indian Chapter, Bangalore.

Ray, T. S., and C. C. Andrews. 1980. Antbutterflies: butterflies that follow army ants to feed on antbird droppings. Science 210:1147–1148.

Reid, J. A. 1941. The thorax of the wingless and short-winged Hymenoptera. Trans. R. Entomol. Soc. Lond. 91:367–446.

Rettenmeyer, C. W. 1960. Behavior, abundance, and host specificity of mites found on Neotropical army ants (Acarina; Formicidae: Dorylinae). Pages 610–612 in H. Strouhal and Max Beier, eds., Proceedings of the Eleventh International Congress of Entomology (Vienna), vol. 1. Organisationskomittee des XI Internationalen Kongresses für Entomologie, Vienna.

Rettenmeyer, C. W. 1961. Observations on the biology and taxonomy of flies found over swarm raids of army ants (Diptera: Tachinidae, Conopidae). Univ. Kans. Sci. Bull. 42:993–1066.

Rettenmeyer, C. W. 1962a. The diversity of arthropods found with Neotropical army ants and observations on the behavior of representative species. Proc. North Cent. Branch Entomol. Soc. Am. 17:14–15.

Rettenmeyer, C. W. 1962b. The behavior of millipeds found with Neotropical army ants. J. Kans. Entomol. Soc. 35:377–384.

Rettenmeyer, C. W. 1962c. Notes on host specificity and behavior of myrmecophilous macrochelid mites. J. Kans. Entomol. Soc. 35:358–360.

Rettenmeyer, C. W. 1963a. The behavior of Thysanura found with army ants. Ann. Entomol. Soc. Am. 56:170–174.

Rettenmeyer, C. W. 1963b. Behavioral studies of army ants. Univ. Kans. Sci. Bull. 44:281–465.

Rettenmeyer, C. W. 1970. Insect mimicry. Annu. Rev. Entomol. 15:43–74.

Rettenmeyer, C. W. 1974. Description of the queen and male with some biological notes on the army ant *Eciton rapax*. Mem. Conn. Entomol. Soc., pp. 291–302.

Rettenmeyer, C. W., and R. D. Akre. 1968. Ectosymbiosis between phorid flies and army ants. Ann. Entomol. Soc. Am. 61:1317–1326.

Rettenmeyer, C. W., R. Chadab-Crepet, M. G. Naumann, and L. Morales. 1983. Comparative foraging by Neotropical army ants. Pages 59–73 in P. Jaisson, ed., Social Insects in the Tropics, vol. 2. Proceedings of the first international symposium organized by the International Union for the Study of Social Insects and the Sociedad Méxicana de Entomologia. Université Paris-Nord.

Rettenmeyer, C. W., H. Topoff, and J. Mirenda. 1978. Queen retinues of army ants. Ann. Entomol. Soc. Am. 71:519–528.

Rettenmeyer, C. W. and J. F. Watkins II. 1978. Polygyny and monogyny in army ants (Hymenoptera: Formicidae). J. Kans. Entomol. Soc. 51:581–591.

Richards, O. W. 1968. Sphaerocerid flies associating with doryline ants, collected by Dr. D. H. Kistner. Trans. R. Entomol. Soc. Lond. 120:183–198.

Rolston, H., III. 1985. Duties to endangered species. BioScience 35:718–726.

Roonwal, M. L. 1975. Plant-pest status of root-eating ant, *Dorylus orientalis*, with notes on taxonomy, distribution and habits (Insecta: Hymenoptera). J. Bombay Nat. Hist. Soc. 72:305–313.

Roosevelt, T. 1910. African Game Trails: An Account of the African Wanderings of an American Hunter-Naturalist. Charles Scribner's Sons, New York.

Rosciszewski, K., and U. Maschwitz. 1994. Prey specialization of army ants of the genus *Aenictus*. Andrias 12. In press.

Rylands, A. B., M. A. O. M. da Cruz, and S. F. Ferrari. 1989. An association between marmosets and army ants in Brazil. J. Trop. Ecol. 5:113–116.

Santschi, F. 1933. Contribution à l'étude des fourmis de l'Afrique tropicale. Bull. Ann. Soc. R. Entomol. Belg. 73:95–108.

Savage, T. S. 1847. On the habits of the "drivers" or visiting ants of West Africa. Trans. R. Entomol. Soc. Lond. 5:1–15.

Savage, T. S. 1849. The driver ants of West Africa. Proc. Acad. Nat. Sci. Phila. 4:195–202.

Schmidt, J. O., M. S. Blum, and W. L. Overal. 1986. Comparative enzymology of venoms from stinging Hymenoptera. Toxicon 24:907–921.

Schneirla, T. C. 1933. Studies on army ants in Panama. J. Comp. Psychol. 15: 267–299.

Schneirla, T. C. 1934. Raiding and other outstanding phenomena in the behavior of army ants. Proc. Natl. Acad. Sci. USA 20:316–321.

Schneirla, T. C. 1938. A theory of army-ant behavior based upon the analysis of activities in a representative species. J. Comp. Psychol. 25:51–90.

Schneirla, T. C. 1940. Further studies on the army-ant behavior pattern: mass organization in the swarm-raiders. J. Comp. Psychol. 29:401–460.

Schneirla, T. C. 1944. The reproductive functions of the army-ant queen as pace-makers of the group behavior pattern. J. N.Y. Entomol. Soc. 52:153–192.

Schneirla, T. C. 1945. The army-ant behavior pattern: nomad-statary relations in the swarmers and the problem of migration. Biol. Bull. (Woods Hole) 88: 166–193.

Schneirla, T. C. 1949. Army-ant life and behavior under dry-season conditions. 3. The course of reproduction and colony behavior. Bull. Am. Mus. Nat. Hist. 94:1–81.

Schneirla, T. C. 1953. The army-ant queen: keystone in a social system. Bull. Union Int. Etude Insectes Soc. 1:29–41.

Schneirla, T. C. 1956. A preliminary survey of colony division and related processes in two species of terrestrial army ants. Insectes Soc. 3:49–69.

Schneirla, T. C. 1957. A comparison of species and genera in the ant subfamily Dorylinae with respect to functional pattern. Insectes Soc. 4:259–298.

Schneirla, T. C. 1958. The behavior and biology of certain Nearctic army ants. Last part of the functional season, southeastern Arizona. Insectes Soc. 5: 215–255.

Schneirla, T. C. 1961. The behavior and biology of certain Nearctic army ants: springtime resurgence of cyclic function—southeastern Arizona. Anim. Behav. 11:583–595.

Schneirla, T. C. 1971. Army Ants: A Study in Social Organization. W. H. Freeman and Company, San Francisco.

Schneirla, T. C., R. Z. Brown, and F. C. Brown. 1954. The bivouac or temporary nest as an adaptive factor in certain terrestrial species of army ants. Ecol. Monogr. 24:269–296.

Schneirla, T. C., R. R. Gianutsos, and B. S. Pasternack. 1968. Comparative allometry in the larval broods of three army-ant genera, and differential growth as related to behavior. Am. Nat. 102:533–554.

Schneirla, T. C., and G. Piel. 1948. The army ant. Sci. Am. 178(6):16–23.

Schneirla, T. C., and A. Y. Reyes. 1966. Raiding and related behaviour in two surface-adapted species of the Old World doryline ant *Aenictus*. Anim. Behav. 14:132–148.

Schneirla, T. C., and A. Y. Reyes. 1969. Emigrations and related behavior in two surface-adapted species of the Old World doryline ant *Aenictus*. Anim. Behav. 17:87–103.

Seevers, C. H. 1965. The systematics, evolution and zoogeography of staphylinid beetles associated with army ants (Coleoptera, Staphylinidae). Fieldiana Zool. 47:137–351.

Sharma, S. K., and O. P. Bohra. 1968. A new record, *Dorylus labiatus* Shuck. (Hymenoptera: Formicidae) as a predator of *Microtermes mycophagus* Desn. (Isoptera: Termitidae). Indian J. Entomol. 30:243.

Shuckard, W. E. 1840. Monograph of the Dorylidae, a family of Hymenoptera Heterogyna. Ann. Nat. Hist. 30:188–201, 258–271, 315–328.

Shyamalanath, S., and J. Forbes. 1980. Digestive system and associated organs in the adult and pupal male doryline ant *Aenictus gracilis* Emery (Hymenoptera: Formicidae). J. N.Y. Entomol. Soc. 88:15–28.

Shyamalanath, S., and J. Forbes. 1983. Anatomy and histology of the male reproductive system in the adult and pupa of the doryline ant *Aenictus gracilis* Emery (Hymenoptera: Formicidae). J. N.Y. Entomol. Soc. 91:377–393.

Smallwood, J. 1982. Nest relocations in ants. Insectes Soc. 29:138–147.

Smith, K. G. V. 1967. The biology and taxonomy of the genus *Stylogaster* Macquart, 1835 (Diptera: Conopidae, Stylogasterinae) in the Ethiopian and Malagasy regions. Trans. R. Entomol. Soc. Lond. 119:47–69.

Smith, K. G. V. 1969. Further data on the oviposition by the genus *Stylogaster* Macquart (Diptera: Conopidae, Stylogasterinae) upon adult calyptrate Diptera associated with ants and animal dung. Proc. R. Entomol. Soc. Lond. 44:35–37.

Smith, M. R. 1942. The legionary ants of the United States belonging to *Eciton* subgenus *Neivamyrmex* Borgmeier. Am. Midl. Nat. 27:537–590.

Smith, M. R. 1965. House-infesting ants of the eastern United States. Tech. Bull. no. 1326. Agricultural Research Service, U.S. Department of Agriculture, Washington D.C.

Snelling, R. R. 1981. Systematics of social Hymenoptera. Pages 369–453 in H. R. Hermann, ed., Social Insects, vol. 2. Academic Press, New York.

Starr, C. K. 1985. Enabling mechanisms in the origin of sociality in the Hymenoptera—the sting's the thing. Ann. Entomol. Soc. Am. 78:836–840.

Stebbing, E. P. 1905. Insect life in India and how to study it. J. Bombay Nat. Hist. Soc. 16:683.

Stephenson, C. 1940. Leiningen versus the ants. Pages 682–703 in A. Gingrich, ed., The Bedside Esquire. Tudor Publishing Company, New York.

Stotz, D. F. 1992. Buff-throated saltator eats army ants. Wilson Bull. 104:373–374.

Strickland, A. H. 1951. The entomology of swollen shoot cacao. I. The insect species involved, with notes on their biology. Bull. Entomol. Res. 41:725–748.

Sudd, J. H. 1959. A note on the behaviour of *Aenictus* (Hym., Formicidae). Entomol. Mon. Mag. 95:262.

Sugiyama, Y., J. Koman, and M. B. Sow. 1988. Ant-catching wands of wild chimpanzees at Bossou, Guinea. Folia Primatol. 51:56–60.

Sumichrast, F. 1868. Notes on the habits of certain species of Mexican Hymenoptera presented to the American Entomological Society. Trans. Am. Entomol. Soc. 2:39–44.

Swainson, W. 1835. A Treatise on the Geography and Classification of Animals. The Cabinet Cyclopedia. Longman, Rees, Orme, Brown, Green, and Longman, and John Taylor, London.

Swynnerton, C. F. M. 1915. Experiments on some carnivorous insects, especially the driver ant *Dorylus;* and with butterflies' eggs as prey. Trans. R. Entomol. Soc. Lond., parts 3, 4, pp. 317–350.

Tafuri, J. F. 1955. Growth and polymorphism in the larva of the army ant (*Eciton (E.) hamatum* Fabricius). J. N.Y. Entomol. Soc. 63:21–41.

Taylor, R. W. 1978. *Nothomyrmecia macrops:* a living-fossil ant rediscovered. Science 201:979–985.

Teles da Silva, M. 1977. Behavior of the army ant *Eciton burchelli* Westwood (Hymenoptera, Formicidae) in the Belem region. II. Bivouacs. Bol. Zool., Univ. São Paulo 2:107–128.

Terayama, M. 1984. A new species of the army ant genus *Aenictus* from Taiwan (Insects; Hymenoptera; Formicidae). Bull. Biogeogr. Soc. Jpn. 39:13–16.

Thorpe, W. H. 1942. Observations on *Stomoxys ochrosoma* Speiser (Diptera, Muscidae) as an associate of army ants (Dorylinae) in East Africa. Proc. R. Entomol. Soc. Lond. 17:38–41.

Topoff, H. R. 1969. A unique predatory association between carabid beetles of the genus *Helluomorphoides* and colonies of the army ant *Neivamyrmex nigrescens.* Psyche 76:375–381.

Topoff, H. 1971. Polymorphism in army ants related to division of labor and colony cyclic behavior. Am. Nat. 105:529–548.

Topoff, H. 1975a. Behavioral changes in the army ant *Neivamyrmex nigrescens* during the nomadic and statary phases. J. N.Y. Entomol. Soc. 33:38–48.

Topoff, H. 1975b. Ants on the march. Nat. Hist. (December): 60–69.

Topoff, H. 1984. Social organization of raiding and emigrations in army ants. Adv. Study Behav. 14:81–126.

Topoff, H., M. Boshes, and W. Trakimas. 1972a. A comparison of trail following between callow and adult workers of the army ant *Neivamyrmex nigrescens* (Formicidae: Dorylinae). Anim. Behav. 20:361–366.

Topoff, H., and K. Lawson. 1979. Orientation of the army ant *Neivamyrmex nigrescens:* integration of chemical and tactile information. Anim. Behav. 27: 429–433.

Topoff, H., K. Lawson, and P. Richards. 1972b. Trail following and its development in the Neotropical army ant genus *Eciton* (Hymenoptera: Formicidae: Dorylinae). Psyche 79:357–364.

Topoff, H., K. Lawson, and P. Richards. 1973. Trail following in two species of the army ant genus *Eciton:* comparison between major and intermediate-sized workers. Ann. Entomol. Soc. Am. 66:109–111.

Topoff, H., and J. Mirenda. 1975. Trail-following by the army ant, *Neivamyrmex nigrescens:* responses of workers to volatile odors. Ann. Entomol. Soc. Am. 68:1044–1046.

Topoff, H., and J. Mirenda. 1978. Precocial behaviour of callow workers of the army ant *Neivamyrmex nigrescens:* importance of stimulation by adults during mass recruitment. Anim. Behav. 26:698–706.

Topoff, H., and J. Mirenda. 1980a. Army ants on the move: relation between food supply and emigration frequency. Science 207:1099–1100.

Topoff, H., and J. Mirenda. 1980b. Army ants do not eat and run: influence of food supply on emigration behaviour in *Neivamyrmex nigrescens.* Anim. Behav. 28:1040–1045.

Topoff, H., J. Mirenda, R. Droual, and S. Herrick. 1980a. Onset of the nomadic phase in the army ant *Neivamyrmex nigrescens* (Cresson) (Hym. Form.): distinguishing between callow and larval excitation by brood substitution. Insectes Soc. 27:175–179.

Topoff, H., J. Mirenda, R. Droual, and S. Herrick. 1980b. Behavioural ecology of mass recruitment in the army ant *Neivamyrmex nigrescens.* Anim. Behav. 28:779–789.

Torgerson, R. L., and R. D. Akre. 1969. Reproductive morphology and behavior of a thysanuran, *Trichatelura manni,* associated with army ants. Ann. Entomol. Soc. Am. 62:1367–1374.

Torgerson, R. L., and R. D. Akre. 1970a. The persistence of army ant chemical trails and their significance in the ecitonine-ecitophile association (Formicidae: Ecitonini). Melanderia 5:1–28.

Torgerson, R. L., and R. D. Akre. 1970b. Interspecific responses to trail and alarm pheromones by New World army ants. J. Kans. Entomol. Soc. 43:395–404.

Townsend, C. H. T. 1897. Contributions from the New Mexico Biological Station. No. 2. On a collection of Diptera from the lowlands of the Rio Nautla, in the state of Vera Cruz. Ann. Mag. Nat. Hist. 19:16–34.

Trivers, R. L., and H. Hare. 1976. Haplodiploidy and the evolution of the social insects. Science 191:249–263.

Tschinkel, W. R. 1987. Fire ant queen longevity and age: estimation by sperm depletion. Ann. Entomol. Soc. Am. 80:263–266.

Tulloch, G. S. 1935. Morphological studies of the thorax of the ant. Entomol. Am. 15:93–131.

van Boven, J. K. A. 1961. Le polymorphisme dans la caste d'ouvrières de fourmi voyageuse: *Dorylus (Anomma) wilverthi* Emery (Hymenoptera: Formicidae). Publ. Natuurhist. Genoot. Limburg 12:36–45.

van Boven, J. K. A. 1967. La femelle de *Dorylus fimbriatus* et *termitarius* (Hymenoptera: Formicidae). Overdruk Natuurhist. Maandblad. 56:55–60.

van Boven, J. K. A., and J. Levieux. 1968. Les Dorylinae de la savane de Lamto (Hymenoptera: Formicidae). Ann. Univ. Abidjan, ser. E. 1:351–358.

Vander Meer, R. K., and D. P. Wojcik. 1982. Chemical mimicry in the myrmecophilous beetle *Myrmecaphodius excavaticollis.* Science 218:806–808.

Vanderplank, F. L. 1960. The bionomics and ecology of the red tree ant, *Oecophylla* sp., and its relationship to the coconut bug *Pseudotheraptus wayi* Brown (Coreidae). J. Anim. Ecol. 29:15–33.

van Lawick–Goodwall, J. 1968. The behaviour of free-living chimpanzees in the Gombe Stream Reserve. Anim. Behav. Monogr. 1:161–311.

Veeresh, G. K. 1990. Pest ants of India and their management. Pages 267–268 in G. K. Veeresh, B. Mallik, and C. A. Viraktamath, eds., Social Insects and the Environment. Proceedings of the Eleventh International Congress of the International Union for the Study of Insects. Oxford and IBH Publishing Company, New Delhi.

Vigah, R., ed. 1977. Accra City Handbook. 1st ed. Public Relations Section of the Accra City Council, Accra, Ghana.

Villet, M. 1989. A syndrome leading to ergatoid queens in ponerine ants (Hymenoptera: Formicidae). J. Nat. Hist. 23:825–832.

Vosseler, J. 1905. Die ostafrikanische Treiberameise (Siafu). Pflanzer 1:289–302.

Vowles, D. M. 1955. The structure and connexions of the corpora pedunculata in bees and ants. Q. J. Microsc. Sci. 96:239–255.

Wallace, A. R. 1878. Tropical Nature and Other Essays. Macmillan, London.

Wang, Y. J., and G. M. Happ. 1974. Larval development during the nomadic phase of a Nearctic army ant, *Neivamyrmex nigrescens* (Cresson) (Hymenoptera: Formicidae). Int. J. Insect Morphol. Embryol. 3:73–86.

Ward, P. S. 1990. The ant subfamily Pseudomyrmecinae (Hymenoptera: Formicidae): generic revision and relationship to other formicids. Syst. Entomol. 15:449–489.

Wasmann, E. 1894. Kritisches Verzeichniss der myrmecophilen und termitophilen Arthropoden. Felix L. Dames, Berlin.

Wasmann, E. 1912. The ants and their guests. Pages 455–475 in Board of Regents of Smithsonian Institution, Annual Report of the Board of Regents of the Smithsonian Institution Showing the Operations, Expenditures, and Condition of the Institution for the Year Ending June 30, 1912. Government Printing Office, Washington, D.C.

Wasmann, E. 1920. Die Gastpflege der Ameisen, ihre biologischen und philosophischen Probleme. Verlag von Gebruder Borntraeger, Berlin.

Wasmann, E. 1925. Die Ameisenmimikry. Ein exakter Beitrag zum Mimikryproblem und zur Theorie der Anpassung. Abh. Theor. Biol. 19:1–164.

Watkins, J. F., II. 1964. Laboratory experiments on the trail following of army ants of the genus *Neivamyrmex* (Formicidae: Dorylinae). J. Kans. Entomol. 37:22–28.

Watkins, J. F., II. 1972. The taxonomy of *Neivamyrmex texanus*, n. sp., *N. nigrescens* and *N. californicus* (Formicidae: Dorylinae), with distribution map and keys to the species of *Neivamyrmex* of the United States. J. Kans. Entomol. Soc. 45:347–372.

Watkins, J. F., II. 1976. The Identification and Distribution of New World Army Ants (Dorylinae: Formicidae). Baylor University Press, Waco, Tex.

Watkins, J. F., II. 1977. The species and subspecies of *Nomamyrmex* (Dorylinae: Formicidae). J. Kans. Entomol. Soc. 50:203–214.

Watkins, J. F., II. 1982. The army ants of Mexico (Hymenoptera: Formicidae: Ecitoninae). J. Kans. Entomol. Soc. 55:197–247.

Watkins, J. F., II. 1985. The identification and distribution of the army ants of the United States of America (Hymenoptera, Formicidae, Ecitoninae). J. Kans. Entomol. Soc. 58:479–502.

Watkins, J. F., II, and T. W. Cole. 1966. The attraction of army ant workers to secretions of their queens. Tex. J. Sci. 18:254–265.

Watkins, J. F., II, F. R. Gehlbach, and R. S. Baldridge. 1967. Ability of the blind snake, *Leptotyphlops dulcis*, to follow pheromone trails of army ants, *Neivamyrmex nigrescens* and *N. opacithorax*. Southwest. Nat. 12:455–462.

Watkins, J. F., II, F. R. Gehlbach, and J. C. Kroll. 1969. Attractant-repellant secretions in blind snakes (*Leptotyphlops dulcis*) and army ants (*Neivamyrmex nigrescens*). Ecology 50:1098–1102.

Watkins, J. F., II, F. R. Gehlbach, and R. W. Plsek. 1972. Behavior of blind snakes (*Leptotyphlops dulcis*) in response to army ant (*Neivamyrmex nigrescens*) raiding columns. Tex. J. Sci. 23:556–557.

Watkins, J. F., II, and C. W. Rettenmeyer. 1967. Effects of army ant queens on the longevity of their workers (Formicidae: Dorylinae). Psyche 74:228–233.

Watts, D. P. 1989. Ant eating behavior of mountain gorillas. Primates 30:121–125.

Way, M. J. 1954. Studies on the life history and ecology of the ant *Oecophylla longinoda* Latreille. Bull. Entomol. Res. 45:93–112.

Way, M. J., and K. C. Khoo. 1992. Role of ants in pest management. Annu. Rev. Entomol. 37:479–503.

Wellman, F. C. 1908. Notes on some Angolan insects of economic or pathologic importance. Entomol. News 19:224–230.

Werringloer, A. 1932. Die Sehorgane und Sehzentren der Dorylinen nebst Untersuchungen über die Facetenaugen der Formiciden. Z. Wiss. Zool. 141: 432–524.

Westwood, J. O. 1840a. An Introduction to the Modern Classification of Insects; Founded on the Natural Habits and Corresponding Organization of the Different Families, vol. 2. Longman, Orme, Brown, Green, and Longman, London.

Westwood, J. O. 1840b. Observations on the genus *Typhlopone*, with descriptions of several exotic species of ants. Ann. Mag. Nat. Hist. 6:81–89.

Wheeler, G. C. 1938. Are ant larvae apodous? Psyche 45:139–145.

Wheeler, G. C. 1943. The larvae of the army ants. Ann. Entomol. Soc. Am. 36: 319–332.

Wheeler, G. C., and J. Wheeler. 1964. The ant larvae of the subfamily Dorylinae: supplement. Entomol. Soc. Wash. 66:129–137.

Wheeler, G. C., and J. Wheeler. 1974. Ant larvae of the subfamily Dorylinae: second supplement (Hymenoptera: Formicidae). J. Kans. Entomol. Soc. 47: 166–172.

Wheeler, G. C., and J. Wheeler. 1984. The larvae of army ants (Hymenoptera: Formicidae): a revision. J. Kans. Entomol. Soc. 57:263–275.

Wheeler, G. C., and J. Wheeler. 1986. Young larvae of *Eciton* (Hymenoptera: Formicidae: Dorylinae). Psyche 93:341–349.

Wheeler, G. C., and J. Wheeler. 1988. The larva of *Leptanilla japonica*, with notes on the genus (Hymenoptera: Formicidae: Leptanillinae). Psyche 95: 185–189.

Wheeler, W. M. 1900. The female of *Eciton sumichrasti* Norton, with some notes on the habits of Texan ecitons. Am. Nat. 34:563–574.

Wheeler, W. M. 1910. Ants: Their Structure, Development and Behavior. Columbia University Press, New York.

Wheeler, W. M. 1911. The ant-colony as an organism. J. Morphol. 22:307–325.

Wheeler, W. M. 1915. On the presence and absence of cocoons among ants, the nest-spinning habits of the larvae and the significance of the black cocoons among certain Australian species. Ann. Entomol. Soc. Am. 8:323–342.

Wheeler, W. M. 1921. Observations on army ants of British Guiana. Proc. Am. Acad. Arts Sci. 56:291–328.

Wheeler, W. M. 1928. The Social Insects: Their Origin and Evolution. Harcourt, New York.

Wheeler, W. M. 1930. Philippine ants of the genus *Aenictus* with descriptions of the females of species. J. N.Y. Entomol. Soc. 38:193–212.

Wheeler, W. M. 1936. Ecological relations of ponerine and other ants to termites. Proc. Am. Acad. Arts Sci. 71:159–243.

Wheeler, W. M., and I. W. Bailey. 1925. The feeding habits of pseudomyrmine and other ants. Trans. Am. Philos. Soc. 22:235–279.

Wheeler, W. M., and W. H. Long. 1901. The males of some Texas ecitons. Am. Nat. 35:157–173.

Whelden, R. M. 1963. Anatomy of adult queen and workers of army ants *Eciton burchelli* Westw. and *E. hamatum* Fabr. (Hymenoptera: Formicidae). J. N.Y. Entomol. Soc. 71:14–30, 90–115, 158–178, 246–261.

Whitacre, D., and D. Ukrain. 1990. Birds at ant swarms in Tikal, Guatamala. Sociobiology 17:467–468.

Whitmore, T. C. 1978. Gaps in the forest canopy. Pages 639–655 in P. B. Tomlinson and M. H. Zimmerman, eds., Tropical Trees as Living Systems. Cambridge University Press, New York.

Williams, E. C., Jr. 1941. An ecological study of the floor fauna of the Panama rain forest. Bull. Chic. Acad. Sci. 6:63–124.

Willis, E. O. 1966. The role of migrant birds at swarms of army ants. Living Bird 5:187–231.

Willis, E. O. 1967. The behavior of bicolored antbirds. Univ. Calif. Publ. Zool. 79:1–132.

Willis, E. O. 1972. The behavior of spotted antbirds. AOU Ornithol. Monogr. 10:1–162.

Willis, E. O. 1981. Diversity in adversity: the behaviors of two subordinate antbirds. Arq. Zool. 30:159–234.

Willis, E. O. 1985a. Surveys of African ant-following birds. Natl. Geogr. Soc. Res. Rep. 21:515–518.

Willis, E. O. 1985b. Behavior and systematics status of gray-headed tanagers (*Trichothraupis penicillata,* Emberizidae). Naturalia 10:113–145.

Willis, E. O. 1985c. East African Turdidae as safari ant followers. Gerfaut 75: 140–153.

Willis, E. O. 1986a. West African thrushes as safari ant followers. Gerfaut 76: 95–108.

Willis, E. O. 1986b. Vireos, wood warblers and warblers as ant followers. Gerfaut 76:177–186.

Willis, E. O. 1986c. Tanagers, finches, and weavers as ant followers. Gerfaut 76:307–316.

Willis, E. O., and Y. Oniki. 1978. Birds and army ants. Annu. Rev. Ecol. Syst. 9:243–263.

Willis, E. O., D. Wechsler, and F. G. Stiles. 1983. Forest-falcons, hawks, and a pygmy-owl as ant followers. Rev. Brasil. Biol. 43:23–28.

Wilson, E. O. 1953. The origin and evolution of polymorphism in ants. Q. Rev. Biol. 28:136–156.

Wilson, E. O. 1958a. The beginnings of nomadic and group-predatory behavior in the ponerine ants. Evolution 12:24–31.

Wilson, E. O. 1958b. Observations on the behavior of the cerapachyine ants. Insectes Soc. 5:129–140.

Wilson, E. O. 1964. The true army ants of the Indo-Australian area (Hymenoptera: Formicidae: Dorylinae). Pac. Insects 6:427–483.

Wilson, E. O. 1971a. The Insect Societies. Harvard University Press, Cambridge.

Wilson, E. O. 1971b. The plight of taxonomy. Ecology 52:741.

Wilson, E. O. 1975. Sociobiology: The New Synthesis. Harvard University Press, Cambridge.

Wilson, E. O. 1978. On Human Nature. Harvard University Press, Cambridge.

Wilson, E. O. 1985a. The sociogenesis of insect colonies. Science 228:1489–1495.

Wilson, E. O. 1985b. Invasion and extinction in the West Indian ant fauna: evidence from the Dominican amber. Science 229:265–267.

Wilson, E. O. 1985c. Ants of the Dominican amber (Hymenoptera: Formicidae). 2. The first fossil army ants. Psyche 92:11–16.

Wilson, E. O. 1985d. The biological diversity crisis. BioScience 35:700–706.

Wilson, E. O. 1987. The earliest known ants: an analysis of the Cretaceous species and an inference concerning their social organization. Paleobiology 13:44–53.

Wilson, E. O. 1988. The current status of ant taxonomy. Pages 3–10 in J. C. Trager, ed., Advances in Myrmecology. E. J. Brill, Leiden.

Wilson, E. O., F. M. Carpenter, and W. L. Brown, Jr. 1967a. The first Mesozoic ants. Science 157:1038–1040.

Wilson, E. O., F. M. Carpenter, and W. L. Brown, Jr. 1967b. The first Mesozoic ants, with the description of a new subfamily. Psyche 74:1–19.

Wilson, E. O., T. Eisner, and B. D. Valentine. 1954. The beetle genus *Paralimulodes* Bruch in North America, with notes on morphology and behavior (Coleoptera: Limulodidae). Psyche 61:154–161.

Wilson, E. O., T. Eisner, G. C. Wheeler, and J. Wheeler. 1956. *Aneuretus simoni* Emery, a major link in ant evolution. Bull. Mus. Comp. Zool. 115:81–99.

Wilson, E. O., and B. Hölldobler. 1988. Dense heterarchies and mass communication as the basis of organization in ant colonies. Trends Ecol. Evol. 3(3):65–68.

Woolley, T. A. 1988. Acarology: Mites and Human Welfare. John Wiley & Sons, New York.

Wygodzinsky, P. 1982. Description of a new species of *Trichatelura* (Insecta, Thysanura, Nicoletiidae) from Ecuador. Sociobiology 7:21–24.

Young, A. M. 1977. Butterflies associated with an army ant swarm raid in Honduras: the "feeding hypothesis" as an alternate explanation. J. Lepid. Soc. 31:190.

Young, A. M. 1979. Attacks by the army ant *Eciton burchelli* on nests of the social paper wasp *Polistes erythrocephalus* in northeastern Costa Rica. J. Kans. Entomol. Soc. 52:759–768.

Young, A. M. 1984. Ithomiine butterflies associated with non-antbird droppings in Costa Rican tropical rain forest. J. Lepid. Soc. 38:61–63.

Author Index

Abe, T., 226, 234
Agosti, D., 21
Akre, R. D.: alarm behavior, 108; beetles, 167, 187, 189, 190–192, 194, 199–200; bristletails, 183, 185; flies, 166–167, 197–198; grooming, 172, 190–191; integrating chemicals, 171, 172; limuloid body form, 180; millipedes, 183; symbionts, 163; trail following, 171; trail substances, 100, 101; trophallaxis, 178
Alexander, B., 50
Alibert, H., 238
Alsop, D. W., 199
Andrews, C. C., 166, 211
Arnold, G., 135
Arnoldi, K. V., 18

Bagneres, A.-G., 53
Bailey, I. W., 82, 83, 127
Baker, H. G., 236
Baldridge, R. S., 79, 101, 214
Barlow, E., 239
Baroni Urbani, C., 10, 12, 40, 57
Barr, D., 15, 16, 19, 21
Bartholomew, G. A., 125–126
Bates, H. W., 30, 43, 201, 205, 246
Beattie, A. J., 53
Beckers, R., 99
Beebe, W., 84, 89, 104, 121–122, 250–251
Bell, R. H. V., 226
Belt, T., 99, 138–139, 206, 246–247
Bequaert, J., 204, 213–215
Bernstein, R. A., 55
Bernstein, S., 55
Bierregaard, R. O., 242
Billen, J.: cell ultrastructure, 57; glands, 34, 53–54, 107; skatole, 110; trail pheromone, 100
Blackwell, M., 196
Blum, M. S., 50–51, 99, 100, 101, 108, 110, 171, 181
Bodot, P., 226, 234
Bohra, O. P., 132
Bolton, B.: Aenictogitini, 40; Cerapachyinae, 10, 34; classification, 8, 35; doryline section, 12; Leptanillinae, 10; phylogeny, 9, 15, 33; subgenital plate, 72

Borgmeier, T., 15, 23, 27, 29, 30, 79, 152, 159, 195, 198
Boshes, M., 102
Bossert, W. H., 225, 236–237
Brauns, J., 128
Brosset, A., 235
Brothers, D. J., 9
Brown, C. A., 110
Brown, F. C., 61, 91
Brown, K. S., Jr., 242
Brown, R. Z., 61, 91
Brown, W. L., Jr.: *Aenictogiton*, 40; alarm pheromone, 107–108, 110; antiquity of ants, 33; army ant behavior in other ants, 37; Cerapachyinae, 10, 34; chromosomes, 57; metapleural glands, 73; wing venation, 23, 71
Bruch, C., 93, 166
Brues, C. T., 176
Bulmer, M. G., 144, 145, 146–147, 157
Burdette, A. W., 34, 74, 75, 156
Burgeon, L., 239
Burton, J. L., 124–125
Buschinger, A., 37
Byatt, A. S., 157
Byrd, J. B., 181

Caetano, F. H., 239
Campione, B. M., 17, 36, 55, 57
Cardinall, A. W., 247
Carlin, N. F., 109–110, 172
Carpenter, F. M., 33
Carpenter, G. D. H., 228, 247–248
Carrel, J. E., 199
Carroll, C. R., 128
Chadab, R., 105–106, 231
Chadab-Crepet, R., 225–226, 231–232, 234
Chao, J. T., 49
Chapin, J. P., 214
Chapman, F. M., 250
Chapman, J. W., 128, 141
Coates-Estrada, R., 208, 209, 242–243
Cohic, F., 45, 122, 212
Cole, T. W., 65, 108–109
Collingwood, C. A., 21
Connell, J. H., 236

Subject Index

aardvark, as army ant predator, 215
aardwolf, as army ant predator, 215
abdominal glands, in ecitonine males, 74
Acadian flycatchers, 207
accessory glands, 74–75; in *Eciton* males, 75; in
   Eciton queens, 66; in *Neivamyrmex* males, 75
Accra (akra), 250
Acritus, 189
aculeate Hymenoptera, 9; in nesting
   associations, 234
aedeagal glands: in *Eciton* males, 73; in
   Neivamyrmex males, 73
aedeagus, 72
Aenictinae, 16–18; furcula, 49
Aenictini: labial palpus of males, 71; labial
   palpus and maxillary palpus of queens, 63;
   mesosoma of workers, 49; waist segmentation, 49
Aenictogitini: *Aenictogiton* as army ants, 39–40;
   affinities with Aenictinae and Dorylinae, 40
Aenictus: absence of cocoons, 85; absence of
   eyes in workers, 47; absence of polymorphism in workers, 57; attacked by *Oecophylla
   smaragdina*, 213; biology, 17–18; cladistic
   analysis, 34; colony fission, 153; column
   raiding, 113; comparative morphology, 34; in
   doryline section, 12; Dufour's gland lining,
   34, 53; exodus flight, 153, 155; geographic
   distribution, 18; as hypogaeic army ants, 36;
   interspecific resource partitioning, 221;
   intraspecific interactions, 158; male
   characteristics, 17; nesting behavior, 88;
   origin, 33; phorid fly associates, 198; queen
   characteristics, 17; search for prey, 121;
   sexual brood production, 148; staphylinid
   beetle symbionts, 195; stinging behavior, 50;
   trail pheromone source, 100; as trophic
   specialists, 128, 132; ventral ganglia, 57;
   worker characteristics, 16–17
Aenictus asantei: emigration behavior, 141;
   nervous system, 55; phorid fly associates,
   198; as presumptively surface-adapted
   species, 36
Aenictus brevicornis, chromosomes (karyotype)
   of, 57
Aenictus eugenii: collection of honeydew, 18;
   tending of *Pseudococcus*, 135

Aenictus gracilis: absence of crop and
   proventriculus in male, 72; alimentary canal
   of male, 72–73; attack of prey, 121; bivouac,
   90; egg-laying capacity, 68; emigration
   behavior, 140–141; functional reproductive
   cycle, 18, 97; as general predator, 17;
   glandular anatomy of male, 73; Malpighian
   tubules of male, 73; nests, 18; rectal pads of
   male, 73; as trophic generalist, 128, 133;
   worker size differential, 45
Aenictus laeviceps: attack of prey, 121; bivouac,
   90; chromosomes (karyotype), 57; column
   raiding, 113; emigration behavior, 140–141;
   functional reproductive cycle, 18, 97; as
   general predator, 17; nests, 18; queen's
   retinue, 139–140; staphylinid symbionts, 195;
   as trophic generalist, 128, 133
Aenictus sp., chromosomes (karyotype) of, 57
Aenigmatiinae, 198; *Aenigmatopoeus*, 179, 198
African mongoose, as army ant predator, 215
African Queen, The, 248
African shrews, 235
African weaver ant, as predator of driver ants,
   212–213
age polyethism. See polyethism
agonistic behavior. See interspecific interactions; intraspecific interactions
Alaopone: geographic distribution, 20; phenetic
   affinities, 19
alarm pheromones: head as source, 107;
   identification pheromone, 108; mandibular
   glands as source, 107; 4-methyl-3-
   heptanone, 107; specificity, 108; and trail
   following, 102–103
alarm response, of *Protopolybia* wasps, 231
Alethe castanea, 206
alimentary canal, of larvae, 83
alitrunk. See mesosoma
alkanes, alkenes, 53
Allomerus octoarticuatus, 234
allometric growth: in *Anomma*, 45; defined, 2,
   43; diphasic, 45; in *Eciton burchelli*, 45; and
   polymorphism, 43–47
altruism, 146
amphibians, as army ant predators, 213
Androeuryops, 201–203; *ecitonis*, 201
Angelaia yepocapa, 232–233

cladistic analysis of army ants, 34
classification: defined, 8; of Formicidae, 8–11; taxonomic history of army ants, 11–16
cleptobionts, 164; *Bengalia depressa*, 203–204
clients, 162–163
close-range olfaction, 110
Cochliomyia hominivorax, 238
cockroaches, 227
cocoons, 85; role in functional reproductive cycle, 95, 97
coefficient of relationship, 145
Coelocraera, 189, 191
colony: defined, 42; features, 2–4
colony fission, 67–68; altruism, 146; bipolar colony organization, 153; callow females, 150; coefficient of relationship, 145; colony growth rate, 147; course of, 148–152; daughter queen, 144; degree of kinship, 147; in *Dorylus (Anomma)*, 153; in *Eciton*, 149–152; eggs, 148–149; and emigration, 153; fertilization, 148–149; Fisher's sex ratio principle, 145–146; and group predation, 144; haplodiploidy, 145; haploid egg production, 148–149; inclusive fitness, 145–148; and kin selection theory, 144–148; latent fission, 149; male eclosion, 150; maternal queen, 146; in *Neivamyrmex nigrescens*, 152; new males, 145; new queens, 145; parental manipulation, 146; parent queen, 145; patrilinial workers, 147; polarization of bivouac, 149; queen choice by workers, 148; queen eclosion, 150; queen pheromones, 148; queen sister, 146; sealing-off reaction, 150; secondary virgin queens, 152; selection pressure, 147; sexual brood production, 148–150; sexual investment ratios, 145; spermatheca, 148–149; spermathecal duct, 149; sperm cells, 148–149; timing, 147; virgin queens, 150–152
colony founding, 67–68, 144. *See also* colony fission
colony odor: acquisition by histerid beetle symbionts, 190; acquisition by myrmecophiles, 110, 172; chemical discriminators, 110; close-range olfaction, 110; cuticular hydrocarbons, 110; in kin selection, 109–110; indole, 110; 3-methylindole, 110; myrmecophile integration, 171–172; in New World army ants, 110; skatole, 110
colony-situation-feedback hypothesis, 68
column raiding: in *Aenictus*, 113; base column, 113; defined, 6; in *Eciton*, 113, 124; in *Neivamyrmex*, 113; in *Nomamyrmex esenbecki*, 113; as plesiomorphic character-state, 36; prey immobilization, 121; raiding strategies, 120; searching for prey, 120–121. *See also* group raiding; swarm raiding
commensals: accidental, 163; defined, 161; preferential, 163. *See also* facultative and obligatory commensals
communication: enabling role in social organization, 94–95; pheromones defined, 99; primer and releaser effects of pheromones, 99; sensory modalities, 99. *See also* alarm pheromones; colony odor; queen

odor; recruitment pheromones; trail pheromones
compound eyes: of *Eciton* workers, 47; of males, 70; of *Neivamyrmex* workers, 47; of New World army ant queens, 63
Conopidae, 201
continental drift, 32–33
convergent evolution. *See* army ant adaptive syndrome
convoluted gland of poison sac, 53
copulation: in *Aenictus*, 157; in *Dorylus*, 157; in *Eciton*, 156–157; fate of male, 79; frequency, 79; in *Neivamyrmex carolinensis*, 156; role of petiolar horns, 63. *See also* mating
corpora allata: of larvae, 84; role in oocyte development, 68
corpora pedunculata, 55
Coxequexomidae, 185
Crematogaster, 212
Cretaceous Period, 32–33
crickets, 227
crop: in *Dorylus (Anomma) molestus*, 60; function, 52; in workers, 52
cuckoos, 207
Cuculidae, 207
cuticular hydrocarbons: and recognition of nestmates, 110; transfer to myrmecophiles, 110, 172. *See also* colony odor
cuticular lipids, 61

daddy longlegs, 227
dealate males, 156–157
defensive posture of workers, 122
Dendrocolaptidae, 207
dessication, 61
detritivores, of middens, 197–198
deutocerebrum, 55–56
developmental convergence, 47
Diapriidae, 195
Dichaetomyia, 201
Dichthadia: geographic distribution, 20; *glaberrina*, 15; phenetic affinities, 19
dichthadiiform ergatogyne, 63
dichthadiigynes, 3; of *Aenictus*, 17. *See also* queens
diet: army ants as carnivores, 128; developmental stage of prey, 134; dietary expansion, 6; of *Dorylus (Alaopone)*, 128; of *Dorylus (Anomma)*, 134; epigaeic–trophic generalist correlation, 128, 132–134; evolution of general predation, 35; expansion as apomorphic character-state, 36; food preference–habitat relationship, 134–135; holometabolous larvae and pupae as prey, 134; honeydew, 135–136; hypogaeic–trophic specialist correlation, 128, 132–134; oligophagy, 128; polyphagy, 128, 134; prey unit size, 134–135; related to group-raiding pattern, 128, 132–134; transport of prey units, 134–135; trophic generalists, 128, 132–134; trophic specialists, 128, 132–134
digestive tract, of workers, 52
dihydrofarnesol, 53
diphasic allometry. *See* allometric growth
Diploeciton: feeding behavior, 192; as obligatory symbiont, 192

maxillary glands: of *Eciton* queen, 65; of
   workers, 53
maxillary palpus: of males, 71; of queens, 63;
   of workers, 47
maxillo-labial complex, 127
Mechanitis, 210–211
median oviduct, 65
media workers, 2. *See also* medium workers
medium workers, in *Eciton burchelli*, 58
Megaponera: with army ant lifeways, 37; *foetens*,
   212; as group-recruitment raiders, 35
Melastomataceae, 234
Melinaea, 211
mesosoma: defined, 47; of *Dorylus* (*Dorylus*)
   helvolus male, 71; of queens, 63, 65; of
   workers, 47–48
metapleural glands: absence in army ant
   males, 73; absence in *Camponotus*, 53; of
   Dorylus, 53; of *Eciton hamatum*, 53; function
   of, 53; of *Neivamyrmex nigrescens*, 53; of
   queens, 65; secretions, 53; of workers, 53
methyl-anthranilate, 100
methyl-branched alkanes, 53
4-methyl-3-heptanone, 107
3-methylindole, 110
Metopininae, 197
middens: *Aenigmatopoeus* as guest, 179;
   arthropod inhabitants of, 93; of bivouacs,
   92–93; detritivores and scavengers, 197–198;
   mite inhabitants, 196; predators, 198–199;
   refuse trails and columns, 93; sphaerocerid
   flies, 198; symbiont inhabitants, 164
midgut, of larvae, 83
migratory nests, 86
military metaphor. *See* army ants: and military
   metaphor
millipedes: as guests, 182–183; as prey, 134
Mimaenictus, 195
mimicry. *See* Wasmannian mimicry
minim workers, in *Eciton burchelli*, 58. *See also*
   minor workers
minor workers, 2
Mischocytarus, 234
mites: classification, 185; dermanyssoid, 176–
   177; as ectoparasites, 196; as myrmecophiles,
   166; phoretic, 176, 185, 187
mkonko, 180–181
molts, 83
monogyny, 67
monoterpene, 53–54
Monte Carlo simulations, 117
morphology. *See* brood morphology; larval
   morphology; male morphology; queen
   morphology; worker morphology
motmot, 208
mountain gorilla, as army ant predator, 215–
   216
mouthparts, of workers, 47
Muscidae, 205
mushroom bodies, 55
Mutilla, 13
Myrmechusa, 195
myrmecoid habitus, 172–177; selected for by
   swarm-following birds, 206; of staphylinid
   beetle, 192

myrmecoid mimicry. *See* Wasmannian mimicry
myrmecophiles: beetles, 167–168; bristletails
   (thysanurans), 183, 185; defined, 162;
   integration index, 163; mites, 166. *See also*
   symbionts and guests
Myrmicinae, as predators of army ant workers, 61

Naked Jungle, The, 252
nasute termite soldiers, 61
natural selection, theory of, 253
Ndoki, 248
Neivamyrmex: accessory glands of males, 75;
   aedeagal gland, 73; biology, 27–28; circadian
   flight times, 79; cladistic analysis, 34;
   cocoons, 85; compound eyes of workers, 47;
   correlation between eye size and circadian
   flight periodicity, 78; in Dominican amber,
   31–32; and ecitophilous diapriid wasps, 195;
   in fossil record, 31–32; geographic
   distribution, 29; imaginal discs in larvae, 46–
   47; intersegmental glandular cells in males,
   74; limulodid beetle symbionts, 187, 189;
   male characteristics, 27; pygidial glands in
   males, 74; phorid fly associates, 198; queen
   characteristics, 27; queen odor, 108–109;
   queen's retinue, 139; reproductive system of
   queen, 65–66; sensory threshold differences
   correlated with worker size, 57; sexual
   brood production, 148; subgenital plate
   gland in males, 74; tergal glands in males,
   74; trail following, 101; trail pheromone
   source, 100; worker characteristics, 27
Neivamyrmex carolinensis: copulation, 156;
   multiple queens, 67; symbiont *Paralimulodes
   wasmanni*, 187, 189
Neivamyrmex harrisi, testicular follicles of, 75
Neivamyrmex nigrescens: attacked by
   Helluomorphoides latitarsis, 199; attack of
   Novomyrmex albisetosus, 229; chemicals
   followed, 100; colony fission, 152; colony size,
   27; column raiding, 113; emigration behavior,
   140, 142–143, 144; emigration influenced by
   food supply, 69, 98; exodus flight, 155;
   functional reproductive cycle, 27–28, 97; larval
   hunger, 138; larval internal anatomy, 83–84;
   larval morphology, 82; and blind snake
   Leptotyphlops dulcis, 181–182; looping
   movements in recruitment, 106; maintaining
   ready attack force, 106–107; mass recruitment,
   107; metapleural glands, 53; nesting behavior,
   88; nomadic phase, 142–144; primary and
   secondary recruiters, 106; recruiting trail
   pheromone, 106; recruitment overrun, 107;
   recruitment to emigrate, 106; sexual brood
   production, 152; skatole production, 110;
   symbiont *Paralimulodes wasmanni*, 187, 189;
   tactile cues in trail following, 104–105; trail
   following in callow workers, 102; trail
   pheromone source, 100–101; worker functions,
   59; worker size differential, 45
Neivamyrmex opacithorax: and ectoparasitic
   fungus, 196; winged males, 152
Neivamyrmex pilosus: column raiding, 113; and
   ectoparasitic fungus, 196; and obligatory
   symbiont, 192